T0235386

New Directions in the Philosophy of Science

Series Editor: **Steven French**, Philosophy, University of Leeds, UK

The philosophy of science is going through exciting times. New and productive relationships are being sought with the history of science. Illuminating and innovative comparisons are being developed between the philosophy of science and the philosophy of art. The role of mathematics in science is being opened up to renewed scrutiny in the light of original case studies. The philosophies of particular sciences are both drawing on and feeding into new work in metaphysics and the relationships between science, metaphysics and the philosophy of science in general are being re-examined and reconfigured.

The intention behind this new series from Palgrave Macmillan is to offer a new, dedicated, publishing forum for the kind of exciting new work in the philosophy of science that embraces novel directions and fresh perspectives.

To this end, our aim is to publish books that address issues in the philosophy of science in the light of these new developments, including those that attempt to initiate a dialogue between various perspectives, offer constructive and insightful critiques, or bring new areas of science under philosophical scrutiny.

Titles include

THE APPLICABILITY OF MATHEMATICS IN SCIENCE
Indispensability and Ontology
Sorin Bangu

PHILOSOPHY OF STEM CELL BIOLOGY
Knowledge in Flesh and Blood
Melinda Fagan

SCIENTIFIC ENQUIRY AND NATURAL KINDS
From Planets to Mallards
P.D. Magnus

COMBINING SCIENCE AND METAPHYSICS
Contemporary Physics, Conceptual Revision and Common Sense
Matteo Morganti

COUNTERFACTUALS AND SCIENTIFIC REALISM
Michael J. Shaffer

MODELS AS MAKE-BELIEVE
Imagination, Fiction and Scientific Representation
Adam Toon

Forthcoming titles include

THE PHILOSOPHY OF EPIDEMIOLOGY
Alex Broadbent

SCIENTIFIC MODELS AND REPRESENTATION
Gabriele Contessa

CAUSATION AND ITS BASIS IN FUNDAMENTAL PHYSICS
Douglas Kutach

ARE SPECIES REAL?
Matthew Slater

THE NATURE OF CLASSIFICATION
John S. Wilkins and Malte C. Ebach

New Directions of the Philosophy of Science
Series Standing Order ISBN 978–0–230–20210–8 (hardcover)
(*outside North America only*)

You can receive future titles in this series as they are published by placing a standing order. Please contact your bookseller or, in case of difficulty, write to us at the address below with your name and address, the title of the series and the ISBN quoted above.

Customer Services Department, Macmillan Distribution Ltd, Houndmills, Basingstoke, Hampshire RG21 6XS, England

Combining Science and Metaphysics

Contemporary Physics, Conceptual Revision and Common Sense

Matteo Morganti
University of Roma TRE, Italy

palgrave
macmillan

© Matteo Morganti 2013

Softcover reprint of the hardcover 1st edition 2013 978-1-137-00268-6

All rights reserved. No reproduction, copy or transmission of this publication may be made without written permission.

No portion of this publication may be reproduced, copied or transmitted save with written permission or in accordance with the provisions of the Copyright, Designs and Patents Act 1988, or under the terms of any licence permitting limited copying issued by the Copyright Licensing Agency, Saffron House, 6–10 Kirby Street, London EC1N 8TS.

Any person who does any unauthorized act in relation to this publication may be liable to criminal prosecution and civil claims for damages.

The author has asserted his right to be identified as the author of this work in accordance with the Copyright, Designs and Patents Act 1988.

First published 2013 by
PALGRAVE MACMILLAN

Palgrave Macmillan in the UK is an imprint of Macmillan Publishers Limited, registered in England, company number 785998, of Houndmills, Basingstoke, Hampshire RG21 6XS.

Palgrave Macmillan in the US is a division of St Martin's Press LLC, 175 Fifth Avenue, New York, NY 10010.

Palgrave Macmillan is the global academic imprint of the above companies and has companies and representatives throughout the world.

Palgrave® and Macmillan® are registered trademarks in the United States, the United Kingdom, Europe and other countries.

ISBN 978-1-349-43391-9   ISBN 978-1-137-00269-3 (eBook)
DOI 10.1057/9781137002693

This book is printed on paper suitable for recycling and made from fully managed and sustained forest sources. Logging, pulping and manufacturing processes are expected to conform to the environmental regulations of the country of origin.

A catalogue record for this book is available from the British Library.

A catalog record for this book is available from the Library of Congress.

Contents

Series Editor's Foreword

The intention behind this series is to offer a new, dedicated, publishing forum for the kind of exciting new work in the philosophy of science that embraces novel directions and fresh perspectives. To this end, our aim is to publish books that address issues in the philosophy of science in the light of these new developments, including those that attempt to initiate a dialogue among various perspectives, offer constructive and insightful critiques, or bring new areas of science under philosophical scrutiny.

Matteo Morganti offers an ambitious and innovative work that clearly meets the aims of the series by addressing the relationship between science and metaphysics, currently very much a 'hot' topic. He attempts to steer a middle course between two well-known views: naturalism, which prioritises science, and what he calls 'conservatism', which shifts the emphasis to philosophical theorising. Morganti offers a 'constructive' naturalism that is both *apriori* and preparatory: On the one hand, it should not be constrained by science, but on the other, it needs science to give it 'substance'. (Excuse the pun!) This framework allows Morganti to propose criteria for metaphysical theory choice, along the lines of a Lewisian cost-benefit analysis.

This overall approach is then applied to three important case studies drawn from the foundations of physics and metaphysics: identity and individuality in quantum theory; substantivalism and relationalism in space-time theory; and composition and the part–whole relationship.

With regard to the first, this has been the subject of considerable discussion, particularly recently. Morganti subjects these discussions to an acute and focussed analysis and concludes that they offer no grounds for giving up on the standard metaphysical notion of individual objects in quantum physics. In particular, he argues for a view of individuality and identity in this context as primitive. However, this kind of ontology fares less well when we move to quantum field theory, and here Morganti lays out various revisionary ontologies with their costs and benefits.

Turning to space–time, Morganti considers the viability of presentism and eternalism, as well as substativalism and relationalism, in the context of modern physics. Applying his framework, he argues that, taken on their own, substantivalism gains the advantage over relationalism in the context of General Relativity. But when presentism and eternalism are brought into the mix, the conservatism associated with the former can be carried over to the relationalist stance, raising its status as a serious option. Again, the overall verdict is not conclusive, but its not meant to be; rather, the options are set out, and their costs laid bare.

The final case examines the metaphysical notions of composition, supervenience and dependence in the context of quantum holism. Here Morganti concludes that although Humean supervenience, for example, must be given up, standard mereological concepts may still be useful, and that the criterion of conservativeness, again, should compel one to adopt a pluralist stance towards fundamental physics, rather than a monistic or structuralist one.

Morganti's book is a bold attempt to re-position the debate over the relationship between metaphysics and science. It offers a provocative and engaging analysis of a range of fundamental issues at the intersection of physics and philosophy. By doing so, it takes the debate forward in new and exciting directions. It is precisely the sort of work that the editorial board and I aim to support through the *New Directions* series, and we are sure that it will be eagerly read by both metaphysicians and philosophers of science.

Steven French
Professor of Philosophy of Science
University of Leeds

Preface and Acknowledgements

This book emerges from and ties together several research threads that I have followed in the last few years. Initially, these focused on the problem of scientific realism and, more generally, on the interplay between physics and philosophy and between philosophy of science and metaphysics. In particular, the study of specific forms of scientific realism prompted me to undertake a more careful analysis of the way in which metaphysical hypotheses play a role within the philosophical analysis of science. The ensuing reflections led me to deal with specific issues in the interpretation of physical theories, and to think more about the general methodology that one should follow when doing scientifically informed metaphysics. Also in view of the recent resurgence in the topic of naturalism in philosophy and metaphysics, putting together some of the ideas that came out of this, I thought, might be of interest for researchers in the field, as well as for advanced undergraduate and postgraduate students.

The book attempts to define a view on the relationship between science and philosophy – more particularly, physics and metaphysics – which satisfies two important requirements at once. On the one hand, that of the naturalists, aiming to make science fundamental for the development of philosophical ideas. On the other, that of the 'conservative', anti-reductionist philosophers who refrain from eliminating philosophical theorising in favour of science. That is, in favour of what would, in effect, be little more than the systematisation and polishing of the results achieved by working scientists that was warmly recommended by the neopositivists. At the same time, the present work proposes a method for actually putting together the indications coming from science and metaphysical hypotheses in specific cases.

The book has five chapters. Chapters 1 and 2 discuss the idea of naturalisation and the notion of naturalised metaphysics, arguing in favour of what may be variously called 'mild', 'liberal' or 'constructive' naturalism. According to constructive naturalism, metaphysics should be intended as a purely *a priori* study and should not be naturalised in the strong sense; that is, by imposing on it the use of experimental methods and scientific concepts and categories, effectively reducing it to science. At the same time, the constructive naturalist also acknowledges that metaphysics is little more than a preparatory enterprise, which delimits the range of what might be the case and necessarily waits for science to say more. This means that science is essential for giving real substance to metaphysics, enabling philosophers to truly say things about the 'structure of reality'. In this sense, metaphysics should be naturalised in the weak sense that it must admit at least a moderate sort of dependence on science. The other aspect of this dynamics is, however, that metaphysics grounds the interpretation of scientific theories, and thus science is, in turn, dependent on philosophy. True, one can be an instrumentalist about scientific theories and consequently consider the philosophical interpretation of science unneeded; and it might also be contended with some plausibility that science, as it is depicted within such an instrumentalist approach, remains much more relevant and useful than pure metaphysics. In a scientific realist setting, however, such asymmetry makes room for a more balanced form of complementarity – or so it will be argued. Criteria for theory choice in naturalised metaphysics are also proposed. These are analogous to those put forward by Lewis in his cost-benefit analysis of philosophical theories in favour of realism about possible worlds. As in the case of Lewis' analysis, it cannot be expected that these criteria can be identified and applied in anything like an objective and completely uncontroversial manner. However, by putting an emphasis on what seem to be the most unquestionable and important of these elements – namely, the fit of one's metaphysical conjecture with the empirical data, and the degree of conservativeness with respect to common sense that said conjecture warrants – the basis for plausible choices, likely to be agreed upon by the majority of philosophers, can be laid.

The rest of the book focuses on three case studies – chosen on the basis not only of personal interest but also of their objective relevance and centrality in philosophical discussion. In each case, the criteria just mentioned are applied with a view to identifying elements of evaluation that can be agreed upon as much as possible, and ways in which the exact extent (if any) to which science intimates a departure from entrenched commonsense beliefs can be determined. Importantly, all these case studies concern physics. This is not due to any anachronistic reductionist presuppositions but rather to personal interest and relevance of the topics in the philosophical literature. (This said, I also believe that there is a sense in which physics is the fundamental science: even non-reductionists will admit that in case of conflict between a physical hypothesis and a conjecture from a science other than physics it is *prima facie* the latter that has to go.) Chapter 3 discusses the problem of identity and individuality in non-relativistic quantum mechanics and (much more briefly) other theories, and suggests a defence of a modest notion of individuality. Something is also said about the recently intense debate on so-called structural realism, in view of the fact that such a position, especially in its most radical version, is directly grounded in specific stances with respect to the interpretation of quantum theory. Chapter 4 deals with space and time and examines how certain traditional philosophical positions (substantivalism and relationism with respect to space and time, presentism and eternalism with respect to time) fare in the face of the available empirical evidence. Here, the tentative conclusion is that more revision is required than in the previous case, even in the context of a constructive naturalism about metaphysics that considers the defence of common sense a priority. At the same time, however, the potential for somewhat different, and more conservative, conclusions is identified in more recent theoretical developments in physics and in the interpretation of it. Chapter 5 closes the book by examining various issues surrounding composition and the part – whole relation. Here, the focus is on how various traditional philosophical theses may or may not survive a science-informed critical scrutiny: for instance, Humean Supervenience, or the idea that constitution is identity.

Overall, I hope that, in addition to a plausible methodology for doing metaphysics, the book will provide useful overviews on certain relevant issues at the boundaries between science and philosophy, and suggestions for positive answers to certain metaphysical questions that will be judged worth taking seriously by people working on these topics.

The writing of the book was made possible by interactions with many people: in particular, Mauro Dorato, Emanuele Rossanese, Mario De Caro and Massimo Marraffa at Roma TRE, Karim Bschir, Claudio Calosi, Angelo Cei, Anjan Chakravartty, Lefteris Farmakis, Steven French, Roman Frigg, Simone Gozzano, Michael Redhead, Samuel Schindler, Mauricio Suarez, Tuomas Tahko and John Worrall as well as audiences in Barcelona, Paris, Madrid, Amsterdam, Pittsburgh, Urbino, Alghero, Milan, Sussex, Bergamo, Konstanz, Bristol, Norwich and Valencia.

1

Metaphysics and Science

1. Philosophy and science

What we now call science appeared in its primitive form as early as the period during which the first Western civilisations flourished. By collecting observational data, elaborating upon them and employing them as a guide for gaining control of things, Babylonians and Egyptians started the study of astronomy, mathematics and medicine. In ancient Greece, knowledge began to be pursued for its own sake and not just for practical purposes. But while it is legitimate to say that the distinction between what one may call the 'pure' and the 'applied' – i.e., between abstract knowledge on the one hand and technical/practical knowledge on the other – was clear already in the Greek culture, the separation between science and philosophy was at that time far from sharply drawn. As a matter of fact, for the Greeks, all those seeking knowledge for its own sake were to be classified as 'philosophers' (lovers of wisdom). And the later Latin term 'scientia' – from which 'science' originates – corresponded to the Greek 'episteme', a concept referring to certain, reliable knowledge in general. Even in the Middle Ages, when philosophy and metaphysics were more precisely defined and systematically developed, those of science and philosophy continued to be more or less interchangeable notions with a rather broad meaning.

In the modern era, however, things changed quite considerably. First of all, what was known as 'natural philosophy' (roughly, the study of nature broadly understood) tended to be based progressively more on experiment and systematic testing of hypotheses. Secondly, the study of reality grounded in experience, as well as the progressive mathematisation of nature, became functional to a general rejection of knowledge sanctioned by dogma and authority. Overall, a distinction, and even opposition, emerged between allegedly 'good', modern philosophy and supposedly 'bad', ancient philosophy – the former emphasising the capability of individual human beings to 'read the book of nature' on their own, and the latter praising the great thinkers of the past and regarding truth as something only a few experts can achieve. The new attitude was most clearly expressed by the British empiricists. Francis Bacon (1526–1626) harshly criticised the Scholastics and, more generally, the Aristotelian tradition for being unable to engage in a truly rational investigation of nature. John Locke (1632–1704), by limiting all our knowledge to two sources, sensation and reflection, excluded the possibility of speculation beyond the facts of experience and consciousness. And this line of thought was taken up by David Hume (1711–1776), who declared it impossible to go beyond experience and systematised in an admirable way this fundamental assumption. Similar ideas were expressed by the thinkers associated with the Enlightenment (around the 18th century) and by positivists in the 19th century, thanks to whom the idea became more and more widespread that the only genuine knowledge comes from experience and lends itself to empirical verification. Auguste Comte (1798–1857), elaborating on Henri de Saint-Simon and Pierre-Simon de Laplace, explicitly presented the scientific method as the supreme guide to knowledge, and science as the most important of human activities. Not surprisingly, around the 1830s, in particular in the work of William Whewell, the word 'scientist' started to connote the systematic natural philosopher following a precise methodology, as opposed to those relying only on *a priori* intuition or on the unsophisticated collection of data of observation. By then, science and philosophy were separated more or less along the lines they are nowadays.

Between the 19th and the 20th century, the praise for science and its method merged with a critique of traditional Aristotelian/Scholastic philosophy (generally labelled 'metaphysics') as based on the improper use of language. George Edward Moore (1873–1958), for example, stated that metaphysical questions arise from misunderstandings due to the bad use of language; and that, consequently, metaphysics is to be regarded as a mere obstacle to the proper inquiry into the nature of things. The anti-metaphysical tendency reached its climax with the development of the neopositivist strand of philosophy (also known as 'logical empiricism') in the first half of the 20th century. Neopositivists put strong emphasis on the verifiability of propositions on the basis of observation. This emphasis implied ruling out the possibility that substantive claims about the world could be formulated without relying on experience, with which the empiricist, positivist and 'proper language' strands of thought came to be fused into one larger system.[1] Proceeding from

Kant's treatment of the 'antinomies or reason', for example, Alfred J. Ayer [1936] argued that, exactly because they are in principle detached from experience and cannot be verified, metaphysical propositions must be deemed meaningless. Similar pronouncements were made, as is well known, by Carnap and others.

On the other hand, the neopositivist project[2] turned out to be a failure for various reasons. Among these, the critique of the sharp distinction between the observable and the theoretical; the Quinean attack on the analytic/synthetic distinction; the Kuhnian and post-Kuhnian views on the history of science and its discontinuities; and, most importantly, the realisation of the fact that verifiability constitutes an inadequate criterion for establishing what counts as meaningful and, thus, what qualifies as a scientific (as opposed to pseudo-scientific or non-scientific) statement. Because of this, not surprisingly, the neopositivist condemnation of metaphysics turned out not to be conclusive. In particular, Karl Popper – while granting that metaphysical statements do not lend themselves to empirical testing – denied that they are meaningless, and instead took them to express ideas and conjectures about reality that it is reasonable to devise and evaluate. Indeed, for Popper, metaphysical views play a constructive role for science itself, at least in heuristically guiding it in its formulation of 'proper' hypotheses and theories.

In general, metaphysics continued to be alive throughout the entire 20th century: first, during the very period in which neopositivism flourished – with thinkers such as Roman Ingarden and Gustav Bergmann;[3] and, afterwards, with the decline of logical empiricism. In the second half of the 20th century, metaphysics even experienced a progressive resurgence. Peter Strawson's "Individuals" [1959] is often regarded as a milestone in this process of revival. There, the distinction was drawn between 'descriptive' and 'revisionary' forms of metaphysics and the relevance of the former – i.e., of metaphysics as the study and clarification of our way of categorising reality in our thought and language and of its necessary preconditions – was defended. Starting from that moment, metaphysics gradually gained popularity, even – if not especially – among the so-called 'analytic' philosophers who can be considered the direct descendants of the neopositivist tradition; and even in its revisionary form, aiming to tell us how we *should* conceive of reality.

The opposition between supporters and detractors of metaphysics, however, continued, and no uncontested winner emerged. In fact, up to today, no consensus has been reached as to what (if any) the relationship between science and metaphysics should be – not even within empiricist[4] circles. In particular, it is true that in recent times, there have been attempts at reconciliation and compromise between science and traditional metaphysical inquiry or, alternatively, attempts to show conclusively that metaphysics is not worth doing, as it is nothing but a remnant of bygone ages not yet illuminated by science and its methodology. And it is true that all these attempts have often been made from a renovated empiricist perspective, now free from the constraints set by the criterion of verifiability. But it is also a fact that a well-defined, shared basis of explicit assumptions and definitions is lacking, and it is therefore not surprising that, so far, the contenders have been unable to truly solve the problematic tension. In particular, metaphysics seems to have been defined only implicitly, via a somewhat vague reference, for instance, to Aristotle and medieval philosophy. Thus, it appears necessary to say something more about science, metaphysics and their mutual relationship. Before undertaking this project, though, it is advisable to introduce another theme, directly relevant for a proper evaluation of the issues we will be interested in.

1.1   Naturalism?

Modern-day empiricists are often self-proclaimed *naturalists*. By 'naturalism', philosophers generally mean the view that 'philosophy must be continuous with the sciences' and 'there is no first philosophy'. That is, great systematic thinking such as that of, say, Kant or Hegel must give way to philosophy as a reflection on our best knowledge of reality, which is one not based on armchair speculation but, rather, on the experience-based methodology distinctive of the sciences. This sounds good. Nonetheless, it remains unclear how exactly this requirement of continuity between science and philosophy and rejection of abstract philosophical constructions should be understood and put into practice. In this sense, as we already mentioned, an imprecise characterisation of metaphysics as 'whatever Aristotle and the Scholastics were doing' seems to be often implicitly at work, which is certainly unsatisfactory.

In practice, the naturalistic credo has led – perhaps unsurprisingly – many once again to seek the elimination of metaphysics along lines similar to those followed and recommended by the logical positivists. In a recent book, for example, Jack Ritchie [2008] defends a sort of 'deflationary naturalism' that invites philosophers to limit themselves to scientific claims and hypotheses as much as possible and ends up effectively eliminating metaphysics. According to Ritchie, the general principle that the naturalistic philosopher has to follow is that of only pursuing philosophical projects that can be carried out through a detailed investigation of science [Ib.; 197]. As an example, Ritchie provides the study of the mutual relationships among the different sciences. His view is that philosophers have often been concerned with such a connection, in some cases promoting reduction and unification (see, e.g., Oppenheim and Putnam [1958]), in other cases endorsing a view of science as a fragmented patchwork of relatively independent parts; but the direct observation and study of actual science, and in particular of actual interdisciplinary work, shows (or at least may show) that neither of the two extreme views is correct, and a more flexible account (or, at any rate, further reflection and analysis) is in order, and this is the direction future philosophical analysis should take.

Independently of the interest of the study recommended by Ritchie in this specific case, his attitude is continuous with respectable philosophical stances that preserve and elaborate on the neopositivist idea that grand systematic philosophical thinking should make room for philosophy as an *a posteriori* reflection on actual science. Indeed, Ritchie overtly acknowledges the closeness of his viewpoint to Fine's [1984] 'natural ontological attitude', the view that science should be taken at face value and no overly ambitious philosophical gloss should be added to it (e.g., realism, antirealism, correspondence theory of truth).[5] However, without entering into a detailed discussion of Fine's views, the worry exists that the deflationary attitude recommended by Ritchie, more or less as in the case of neopositivism, sacrifices too much in proportion to the actual strength of the arguments provided in its support. One first obvious thought is that a deflationary view of the sort just described runs the risk of not taking adequately into account the well-known recommendation that metaphysics be preserved in its traditional form because, in that form, it is part and parcel of a lot of what goes on in actual scientific theory-making. More generally, the risk is that one presupposes a sharp divide between science and metaphysics which either does not exist or can only be drawn in ways – yet to be identified – which, as a matter of fact, do not serve one's intended purposes. In connection to this, it is interesting to notice (following Friedman [2001; 12–13]) that supposedly paradigmatic figures of scientists such as Helmholtz and even members-to-be of the Vienna Circle such as the young Schlick did not see 'scientific philosophy' as necessarily grounded on this sort of deflationist stance. Schlick, for instance, argued at one point that we must begin with special problems of the special sciences but then move up "to the ultimate attainable principles ... which, because of their generality, no longer belong to any special science, but rather lie beyond them ... in philosophy" [1978; 335]. Be this as it may, the very definition of naturalism that Ritchie starts from (i.e., the abovementioned requirement of *continuity* between science and metaphysics) leaves room in conceptual space for alternatives that should be evaluated with greater care. As a matter of fact, Ritchie himself acknowledges the possibility of a more ambitious form of naturalism about metaphysics. He identifies it with what he calls 'constructive methodological naturalism', which he characterises on the basis of its appeal to inference to the best explanation to justify metaphysical conclusions. However, that the use of inference to the best explanation suffices in itself for delineating an interesting form of metaphysics, and one that should appear relevant from the perspective of the scientifically-minded philosopher, is unlikely. Is not the use of abductive method the distinctive feature of *any* attempt (be it scientific or philosophical) to move from the known to the conjectural? Not surprisingly, after briefly examining this option, Ritchie sets aside constructive methodological naturalism in favour of his own, philosophically less ambitious, variety of naturalism. Equally unsurprisingly, not everyone will be satisfied with his assessment of forms of naturalism alternative to the deflationary variety.

Another version of naturalised metaphysics is defended and articulated by Ladyman and Ross [2007]. These authors argue against what they call 'neo-Scholastic metaphysics', which they connote as an activity that puts forward abstract claims and hypotheses that only pay lip service to science[6] and are, in fact, grounded almost exclusively in commonsense intuition. As an alternative, Ladyman and Ross defend a conception of metaphysics as the search for, and promotion of, unification among scientific theories on the basis of physics. (They contend that such unification is best achieved by endorsing a specific revisionary metaphysical viewpoint – more on this in later sections.) Naturalised metaphysics, they suggest, should follow two basic principles. The 'Principle of Naturalistic Closure':

> Any new metaphysical claim that is to be taken seriously at time *t* should be motivated by, and only by, the service it would perform, if true, in showing how two or more specific scientific hypotheses, at least one of which is drawn from fundamental physics, jointly explain more than the sum of what is explained by the two hypotheses taken separately. [Ib.; 37]

And the 'Primacy of Physics Constraint':

> Special science hypotheses that conflict with fundamental physics, or such consensus as there is in fundamental physics, should be

rejected for that reason alone. Fundamental physical hypotheses are not symmetrically hostage to the conclusions of the special sciences. This, we claim, is a regulative principle in current science, and it should be respected by naturalistic metaphysicians. [Ib.; 44]

The second principle appears reasonable, especially given the detailed and plausible explication provided by Ladyman and Ross of the sense in which physics is prior to the other sciences, so let us focus on the first. Why is it that taking science seriously should compel one to endorse Ladyman and Ross' view of metaphysics as only useful when securing explanatory unification? Could not metaphysics and science enter into a virtuous mutual relationship without producing any explanatory unification between scientific hypotheses? For instance (anticipating a later theme), would Ladyman and Ross have been ready to give up on their structuralist metaphysics – according to which reality is ultimately analysable in terms of relations – had it turned out to apply only to the quantum mechanical domain (from a consideration of which it originates)?[7] Ladyman and Ross point out [Ib.; 65] that their use of the notion of metaphysics is idiosyncratic. The fact remains, though, that they present *their* form of metaphysics as *the* form of metaphysics that the naturalist should find most sensible; but they have not made a compelling case in favour of the claim that what they present is the only (or, at any rate, the most plausible) non-eliminativist alternative to whatever qualifies as bad, neo-Scholastic metaphysics in their sense.[8]

Speaking of which, let us get back for a moment to Ladyman and Ross' rejection of neo-Scholastic metaphysics. As we have seen, one element of their criticism is that neo-Scholastic metaphysics does not really take the best available science seriously. This is ambiguous: do contemporary analytic metaphysicians crucially rely on a caricature of science to reach their conclusions, or are simplified systems sufficient for their arguments and purposes? If the latter, where is the problem, exactly? As we will see in what follows, there are a lot more shades in the relationship between science and metaphysics, and in the idea of 'taking science seriously', than Ladyman and Ross seem to acknowledge. Ignoring the details in favour of sweeping generalisations is likely to be a bad choice here. Moving on to the second criticism levelled by Ladyman and Ross against neo-Scholastic metaphysics, concerning intuition – it, too, seems to be based on ambiguity. Ladyman and Ross conflate a) the employment of premises lacking explicit argument in support of them, and b) the stubborn defence of commonsense. Since non-explicitly-argued-for premises are inevitable in every piece of reasoning, and it is an unquestionable matter of fact that metaphysicians are simply not (or, at any rate, need not be) interested in defending our naïve view of the world come what may, it is plausible to think that this part of Ladyman and Ross' argument has no real force, as they mistake a rhetorical/argumentative way of proceeding for a statement of intent (on this, see Dorr [2010]).[9] (It is also worth mentioning that the role of intuitions in physics is perhaps less clearly negligible than Ladyman and Ross assume it to be. For an interesting study, see Tallant [forthcoming]).

In general, it looks as though what it means to implement a form of naturalism in philosophy remains far from clear, and certainly requires further discussion. Unquestionably, this holds in the specific case of metaphysics, with respect to which a lot of present-day scepticism appears to be rooted in little more than neopositivist prejudice – albeit with the criterion of verifiability having now been substituted by a more acceptable, but also considerably more vague, emphasis on the priority of science when it comes to seeking knowledge of the world.[10]

This goes to confirm that, in order to say something more about science and metaphysics, it is necessary to be more precise as regards what exactly is distinctive of each of them, and where the difference between the two specifically lies. This is the task we now turn to. After that, we will envisage a form of naturalism about metaphysics alternative to those currently on offer.[11]

2.   Where's the difference?

Let us start from the basic definitions. Roughly speaking, science can be intended as *the systematic enterprise of*

i) Gathering information about the world through observation and experiment;
ii) Formulating testable hypotheses and full-blown theories on the basis of that information (and its elaboration through mathematics) – so obtaining instrumentally useful knowledge of the relevant domain.[12]

As for metaphysics – as readers may have expected – things are more complicated. Historically, the label 'metaphysics' is little more than the by-product of circumstance. Aristotle (384–322 B.C.) produced a number of works which together were called the *Physics* – essentially, treatises aimed to uncover the fundamental principles of change and motion. Later scholars working on the systematisation and organisation of Aristotle's writings (most notably, Andronicus of Rhodes around three hundred years after Aristotle's death) placed right after the *Physics* another set of writings which was named "those (works) that come after the (works about) physics": 'ta meta ta physika' – from which the word 'metaphysics'. This term, of course, pointed to the fact that the subject matter of those texts concerned things that underlie the empirical knowledge of the physical and are prior to it, and not just to a contingent fact regarding editorial organisation.[13] That of metaphysics, however, was by no means regarded as a self-explanatory notion. First of all, metaphysics was not a unitary subject. Aristotle's *Metaphysics* was divided into three parts: 'Ontology', 'Theology' and 'Universal Science'. Theology was the study of God (or the gods), while universal science was the study of so-called 'first principles', that is, of the elementary laws of logic. Ontology was, instead, the discipline that was later defined as the science of 'being *qua* being', that is, of what exists in its most general traits.

It is clear that what we call metaphysics nowadays is closest to Aristotle's ontology, since theology and logic are now undoubtedly distinct from, although related to, what we take to be metaphysics. However, identifying metaphysics with ontology as the Aristotelian study of being is not correct. After all, the terms 'metaphysics' and 'ontology' are often taken to denote different activities and, *prima facie*, this linguistic differentiation seems to point to a meaningful conceptual distinction. Indeed, it is agreed, by and large, that ontology as the study of the categories of entities that exists[14] is but a specific branch of metaphysics. Of course, the open question remains of what exactly the term 'metaphysics' should be intended to mean besides whatever is conveyed by the vague term 'being in itself'.

Our primary aim is, then, to define metaphysics in the most precise and plausible manner, and in such a way that it is (if possible) properly and neatly differentiated from both ontology and science – hopefully, in the latter case, not in a relationship of competition and mutual exclusion. In the rest of this section, we will consider various options for doing so.

2.1 Scope

A relatively safe definition of metaphysics is its most commonly accepted one: that is, that according to which metaphysics is *the study of the fundamental structure of reality in general',* of the *'ultimate nature of everything that exists.* This means that metaphysics presents itself as the most wide-ranging study of reality, not limited to any particular subdomain of it. Indeed, it might appear uncontroversial that if one studies the basic features of *everything*, one is *ipso facto* doing something pretty different from what scientists do. For, scientists, it would seem, work in specific fields and aim to formulate testable hypotheses about sub-regions of the universe – where the boundaries of each such sub-region may be questionable and fuzzy, but are surely there as something that defines a proper subset of what exists.

However, this is not so. As a matter of fact, by itself a definition in terms of scope *cannot* suffice for demarcating between science and metaphysics.

First of all, there is a scientific discipline that has as its specific object of study the universe as a whole: cosmology. By definition, there is nothing that exists as a material entity and falls outside of the scope of inquiry of cosmology. One may reply that cosmologists do not aim to account for absolutely everything but 'just' for the most comprehensive physical system; and that other domains of inquiry exist that are not explained (nor considered as something to be explained) by cosmologists. Here, more should be said about these other domains of inquiry: does the reply refer to non-material entities such as souls or numbers? If so, granting that metaphysicians deal (or at least, may deal) also with entities of this latter kind, there indeed seems to be a difference between cosmology and metaphysics. Does it refer to non-actual entities? If so, in spite of appearances, again the domain of metaphysics seems to be distinct from that of cosmology.

First of all, let us assume here that the scope of metaphysics, or, at any rate, of the sort of metaphysics that should be deemed inter-esting, relevant or worth pursuing, is restricted to existing material entities. (Alternatively, let us suppose that materialism/physicalism is correct and that everything non-material/non-physical can be reduced to something material/physical.[15] This is surely an innocent assumption to be made here, as in any case even if metaphysics were not only concerned with material entities, one should still explain why it is called 'metaphysics' rather than 'science' when it deals with those.) Even then, the criticism to the proposed definition of metaphysics stands. Because, no matter whether it is plausible to believe in its possibility and a sensible aim to try to achieve it, a so-called 'Theory of Everything' formulated in the vocabulary of a given special science (most likely physics) is conceivable, and in fact often discussed.[16] And what people normally intend when talking of a Theory of Everything is a putative product of theoretical *physics* that fully explains and links together all known phenomena, and, ideally, has predictive power with respect to the outcome of any experiment that could be carried out in principle – where, crucially, this explanatory and predictive power is rooted in statements and procedures which are typical of (physical) science. It follows from this conception of what a Theory of Everything is that such a theory (at least as we understand it now) would be an entirely scientific theory. In particular, it would be a theory that uses the concepts and methods distinctive of one specific science, and differs from usual scientific theories in the same domain only with respect to explanatory power and scope of application. In light of this, it should appear clear that simply calling a physical Theory of Everything a piece of metaphysics would just be a terminological choice and would not mirror a proper characterisation of metaphysics but only (if anything) a contingent fact about contemporary science.[17]

The defining feature(s) of metaphysics, then, must lie somewhere else.

2.2 Questions and answers

Perhaps emphasis should be put not on the fact that metaphysics studies 'everything that exists', but that it studies the 'fundamental structure' of it. That is, perhaps the correct characterisation of metaphysics is in terms of the questions that it asks – and claims to be able to answer – rather than the scope of such questions.

This is what several anti-metaphysical philosophers seem to think. Some of them point at (allegedly) typically metaphysical questions and acknowledge their peculiar nature. But they do so with a view to arguing that it is manifestly pointless to look for a compelling answer to those questions and that, consequently, these should be set aside as a waste of time – the same holding, obviously enough, for (neo-Scholastic) metaphysics in general. Putnam [2004], for instance, takes the question "How many objects are there in a mini-world with exactly three point-particles?" to be a paradigmatic metaphysical question, and then goes on to attack metaphysics by arguing that, like analogous ones, this question has neither a clear answer nor a way of being tackled properly. After all, what could possibly give us an indication as to whether or not the sum of two things itself counts as 'one' thing? How can anything informative be added to the initial description of Putnam's mini-world? Putnam does not hesitate to generalise and conclude that analytic metaphysics should not be pursued at all. In a similar vein, Van Fraassen [2002] discusses the question "Does the world exist?" and ends up contending that it has been shown beyond doubt that 'metaphysics is dead'. This line of argument, however, needs some fleshing out.

To begin with, when one looks at the historical development of the subject, one can see that there is no obvious continuity in terms of issues and questions in metaphysics. Van Inwagen [2012], for example, argues that there is a clear difference between the 'old' metaphysics studied by Aristotle and medieval thinkers, and the 'new', post-medieval metaphysics. According to Van Inwagen, for instance, the study of substance and of the categories of being belongs to old metaphysics, while questions concerning modality the relationship between the mental and the physical and the constitution of material objects are typical of the new metaphysics. Independently of its details, Van Inwagen's 'classification' allows one to see that, *en route* to a critique of metaphysics, the examples provided by Putnam and Van Fraassen (henceforth, PVF) should be supplemented with at least two things:

i) A sufficiently precise and agreed upon indication of what questions are to be regarded as metaphysical nowadays (presumably, PVF would agree that not all questions that seem to lack a precise and objective 'answering strategy' count as metaphysical);[18]
ii) An argument to the effect that that what is the case for the presented 'paradigms' is also characteristic of all, or of the majority of, the questions identified as metaphysical.

With respect to (i), PVF seem to assume that the domain of metaphysics has already been defined in a sufficiently clear-cut way at least by the practice of metaphysicians, which is, of course, something one may quarrel with. Suppose, however, that the first task is accomplished and at least a minimal, non-empty set of questions that certainly qualify as metaphysical is identified. The issue remains open (ii) of *what exactly* it is, if anything, that makes those questions 'of the same kind', and is, consequently, indicative of the fact that analytic metaphysics is a bad intellectual enterprise. Just answering 'the fact that they lack an answer in principle' would be question-begging.

The immediate thought is that, unlike scientific hypotheses, the conjectures one may employ to answer metaphysical questions are all equally (im)plausible because remote from the actual world and completely untestable. That is, that in metaphysics one only obtains answers in the form of what one may call 'pure hypotheses', that is, conjectures about things that cannot possibly be observed and tested against empirical data. Indeed, Van Fraassen [2002; 12–16] expresses this 'remoteness objection' quite clearly as he argues that the separatedness of metaphysical conjectures from empirical considerations makes them not meaningless but certainly useless. He argues, in particular, that scientific hypotheses can be (and are constantly) put to severe tests, and this may well result in their being falsified; but, far from constituting a problem for science, this bears witness to its strength and significance. Metaphysics instead, says Van Fraassen, seeks truth but is never in a position to establish, or even probe, whether what it says is actually true or false on the basis of available evidence. Therefore, even though both metaphysics and science employ inference to the best explanation, and so move from the known to the conjectural, the former but not the latter, according to Van Fraassen, turns out to be a merely formal exercise.

At a first glance, there seems to be something to this argument. But how is a principled distinction to be drawn between pure hypotheses in the above sense and hypotheses that are not hopelessly remote from the empirical domain? That is, how is Van Fraassen's generalisation to be justified exactly? Recall that using testability and verifiability to divide science and non-science already turned out to be a failure in the past.[19]

Even independently of the formal considerations that were brought to bear against the criteria of verifiability proposed by the neopositivists, mere testability will not do for even simpler reasons: claims about the sphericity of the Earth or about craters on the dark side of the moon, for instance, were not testable via direct observation when they were first formulated[20] but are clearly so testable now. And it must be acknowledged that the same may well occur for hypotheses that seem to count as pure nowadays, and even for less direct forms of testability: of course, the denial of this possibility as a matter of principle and/or on the basis of mere intuition would not be in line with a broadly understood empiricism.[21]

However, Van Fraassen seems to have a point when he claims [2002; 24] that, in the case of metaphysical questions and hypotheses, the basic concepts and notions are irreducibly context-dependent; that is, they are indissolubly dependent on assumptions and definitions that cannot be objectively evaluated and compared, no matter how much evidence is brought to bear. (This is clearly reminiscent of Carnap's talk of *internal* as opposed to *external* questions.) As a matter of fact, the questions considered by PVF (to recall them: "How many objects are there in a mini-world with exactly three point-particles?" and "Does the world exist?") do seem to have something unconvincing about them. Thus, instead of insisting on the 'no sharp boundary' line of argument, one might prefer to attempt to elaborate on this point. Perhaps it could even be explicitly used as a ground for a defence of metaphysics.

In a sense, what we just discussed is nothing new, as Van Fraassen's remoteness objection basically reiterates the well-known Kantian point about those assertions that "lay claim to insight into what is beyond the field of all possible experiences" [*Critique of Pure Reason*; A425/B453] – that is, about those assertions that constitute the four 'antinomies of reason'. What needs to be established here is whether reasons can be found for resisting, in spite of the remoteness of metaphysics from the empirical domain, PVF's eliminativist conclusions with respect to such assertions and questions. If the answer to this turned out to be affirmative, of course, a more positive characterisation of metaphysics should then be provided. In this connection, two points need to be discussed.

The first aspect concerns the notion of empirical relevance. Consider, for instance, a metaphysical category that many regard as a useless relic of past philosophy: that of universals. And compare the items it contains with typical scientific hypothetical entities: say, gravitons. In the case of universals, the enemy of metaphysics may suggest, there is no way of giving empirical support to claims for or against them. How could we establish, on the basis of experience, whether two things with a property in common literally share numerically the same entity? Van Fraassen's remoteness objection appears based exactly on doubts of this sort. Gravitons, instead, while currently being entirely hypothetical entities, are defined on the basis of a well-defined model (the standard model of elementary particles based on quantum field theory) in such a way that it is pretty clear what would count as evidence confirming their existence. So much so, that large – and very expensive! – particle accelerators can be (and have been) built to try to obtain such confirming evidence in the past: think of the recent results lending empirical support to the conjectured existence of Higgs bosons. To be sure, independently of whether or not similar results could be obtained for gravitons, nobody would do the same to detect universals. In general, it could be insisted that we progressively come to 'see more' in science but not in metaphysics: consider, for example, the abovementioned idea that the Earth is spherical rather than flat, and how it has switched from the status of little more than mere conjecture in, say, the 6th century B.C., to that of a directly observable truth in the 20th century – thanks to space shuttles and the like. Undeniably, the same progress has not been achieved with respect to universals (*qua* repeatable properties).

The above seems essentially right, thus legitimating the requests made by past and contemporary empiricists independently of the technical and conceptual shortcomings of criteria of verifiability as these have actually been formulated. However, this is still *not* sufficient for supporting Van Fraassen's conclusion that metaphysics is a merely formal game. For, it does *not* conclusively show that one's views, say, on the nature of properties do not make *any* empirical difference or, better, any difference to our knowledge and understanding of the empirical domain. Consider, for instance, the following example (which we will discuss in much more detail in Chapter 3). The issue concerning the (non-)individuality of quantum particles relies more or less implicitly on assumptions about the nature of properties and individuation. These assumptions can only be spelled out in philosophical, not scientific, terms. In continuity with the example used a moment ago, consider properties, and the infamous 'problem of universals' in metaphysics. Realists about universals who do not also postulate substrata/bare particulars are compelled to endorse Leibniz's principle of the Identity of the Indiscernibles.[22] But the latter is exactly what seems to be put into doubt by quantum mechanics. This might be read as indicating that, with the advent of quantum physics, realism about universals has become an empirically relevant thesis – at least in the sense that the hypothesis that universals exist is now *indirectly testable*!

Of course, supporters of universals postulate them in order to fulfil a number of explanatory roles which need not be affected by, nor necessarily related to, empirical considerations. The point is that the latter, at least in the form of interpretation-related considerations, become relevant alongside other evaluative criteria that typically lead theory-choice in metaphysics (a more detailed discussion of which will be offered in Chapter 2).

Elaborating on this, one could thus *agree* with what PVF say about metaphysical questions while *resisting* their eliminativist recommendations. In particular, one may suggest that any hypothesis one puts forward about reality should be required to make *some sort* of difference when it comes to evaluating the data of observation and the results of our experiments. Some of them will perhaps be confirmed or refuted directly (obviously enough, in a defeasible manner). But others may instead turn out to be such that their truth or falsity cannot be ascertained, and yet assuming them affects our views about other matters, as in the above example concerning identity and individuality in quantum mechanics. Although the dividing line is probably (perhaps inevitably) blurred, this will be deemed sufficient for present purposes. Consequently, it will be assumed that metaphysical questions and the hypotheses they allow as possible answers can be distinguished in principle from scientific questions and hypotheses on the following basis: that the former (thanks to their generality, the concepts they involve and the form of reasoning they require – more on this shortly) are exactly the sort of things that can (in fact are needed to) ground one's *interpretation* of the scientific theories based on the latter.[23]

Van Fraassen does take this possible understanding of metaphysics as available and worth considering. He, however, also thinks that the fact that it grounds interpretation cannot be a reason for considering metaphysics valuable, for "metaphysicians interpret what we initially understand into something hardly anyone understands" [2002; 3]. But this claim of Van Fraassen's is doubly unwarranted: both in taking science to be already understood, and in taking metaphysical notions and hypotheses to be (necessarily) obscure. On the one hand, non-interpreted science is not as much 'understood' as 'usable'. For, in what sense does one understand a more or less formal apparatus, or a family of models as a representation of reality? At best, in such a case one has merely practical knowledge, but that hardly deserves to be classified as a form of understanding. Slightly differently put, there might be a sense in which scientific theories explain and provide understanding on their own, and maybe even have a 'natural interpretation'. But the deeper meaning of such theories can only come to the surface through sophisticated philosophical analysis. On the other hand, once it is aptly defined and developed (mainly if not exclusively, as we are arguing, with a view to interpreting science), metaphysics is far from being incomprehensible, as it simply systematises into appropriate categories and concepts questions that are importantly continuous with those of science on one side, and of common sense on the other. (This will become clearer shortly, as a more specific characterisation of metaphysics will be provided.) The upshot of this is that instrumentalists can coherently reject metaphysics, but only because they do not truly seek understanding and knowledge, and are instead happy to aim at merely applicative success, obtained via the application of practical abilities. The more ambitious sense of knowledge presupposed by a generally realist attitude, instead, requires that space be given to interpretation and thus, we are suggesting, to metaphysics intended in a non-reductive fashion.

Getting back to the PVF sort of objection, then, employing hypothetical posits to interpret science certainly counts as a form of empirical relevance which does not involve a conclusive pronouncement with respect to the truth or falsity of those hypotheses. With which, the objection appears decisively deflated.

The consideration of another aspect of the issue will allow us to further qualify the definition of metaphysics (at least the one that will be accepted here). We have at least implicitly acknowledged in the foregoing that certain questions are indeed 'bad' from a perspective according to which responsiveness to the input coming from experience is essential. But why is it so? The suggestion in what follows will be that it is because they belong to the *wrong sort* of metaphysics, i.e., they emerge as relevant questions only in the context of a conception of metaphysics that can (and should) be set aside.

In particular, the immediate target of PVF's criticisms are *existence* questions that, *as such*, problematise something that many would regard as unproblematic and/or self-evident. But such questions can be taken to be paradigmatic of metaphysics as a whole only if metaphysics as a whole is intended as primarily concerned with existential questions. This indicates that it is a working presupposition for PVF that metaphysics has to be conceived of on the basis of Quine's famous reflections 'on what there is' and his dispute with Carnap. As is well known, Quine and Carnap debated over whether existential commitment should only be made within a specific linguistic and conceptual framework (Carnap [1950]), or should follow from a careful observation of what entities our best theories in the various domains of inquiry rely upon and cannot eliminate,[24] in such a way that the ensuing conclusions are regarded as valid in general (Quine [1951]). In this sense, with Quine, ontological questions as external questions in Carnap's sense gained new dignity.[25]

The debate itself, and the Quinean viewpoint in particular, thus took for granted that metaphysics is about what exists, and about answering all questions relevant for determining exactly what things we should believe to be out there. Indeed, such a conception directly leads one to take seriously the specific questions considered by PVF and, more generally, a whole lot of questions that may legitimately be regarded as empty, or at least lacking an objective answer, from a scientific point of view: for instance, the infamous question of whether there is only a table there, or just a set of particles arranged table-wise, or both. However, it is perfectly possible, and in fact advisable, for metaphysicians to reject the Quinean definition of metaphysics. For instance, Lowe [2011] convincingly distinguishes between bad (Quinean) metaphysics and good metaphysics, and Schaffer [2009] compellingly argues that defending metaphysics on the basis of the Quinean (as opposed to the Carnapian) view is a non-starter.[26] Schaffer, in particular, overtly recognises that existence questions are, more often than not, empty or at least uninteresting; that they demand answers that can only be given on the basis of subjective assumptions and premises; and that they consider problematic things that are, at root, not problematic at all. As a matter of fact, argues Schaffer, what exists is not the issue, but the self-evident starting point. At most, that is, ontology intended in the traditional Quinean fashion can only be an instrument for a more fundamental sort of inquiry.[27] The alternative that Schaffer recommends is that according to which metaphysics is primarily a study of grounding relations, that is, of what is more fundamental than what and how the structure of the world is determined by such priority and dependence relations. Lowe, instead, puts particular emphasis on the fact that metaphysics deals with possibilities. We will look at these ideas in more detail in the next chapter. For the time being, it is sufficient that there is an alternative to the 'canonical' understanding of metaphysics which, as we have seen, is in fact open to criticism.

In light of the foregoing discussion, it looks as though drawing a distinction between science and metaphysics in terms of the nature of the questions asked and of the hypotheses formulated – although it requires some caution – appears both possible and advisable. Doing so, however, doesn't give us a compelling reason for endorsing eliminativism about metaphysics (unless, we argued, one assumes an instrumentalist stance towards science to begin with). On the contrary, it offers us a hint as to how 'proper' metaphysics may be characterised and connected to science. In order to start defining in more detail such metaphysics, before closing this first chapter we shall consider one last possible way of defining metaphysics and demarcating between it and science.

2.3 Concepts, categories and methods

The reason why a physical Theory of Everything does not count as metaphysical, we have seen, is that such a theory is to be formulated in terms of *concepts and categories* that are those commonly used by previous physics, or likely to be continuous in the important respects with those used by previous physics. The same holds, more generally, for every scientific theory and for every special science. Historically, instead, metaphysics has consistently used concepts and categories that are different from those of the special sciences and intended to have the most general applicability: for instance, those of substance, object, property, universal, or causation. One may thus plausibly suggest that it is the use of these concepts and categories, supposed to be applicable independently of the specific part or domain of reality that one is observing, that grounds the status of metaphysics as an autonomous discipline. This is clearly relevant in connection with the suggestion made in the previous section: If the concepts and categories employed by metaphysics are different from, and more fundamental than, those of the special sciences, it does not come as a surprise that metaphysical questions and hypotheses can be used to make sense of, i.e., interpret, science. And it becomes also plausible that these concepts and categories play a crucial role in answering not existential questions but questions of other types – possibly having to do with ontological grounding and the dependence/priority structure of reality.

But does this not also mean that, contrary to what seemed to emerge in the previous section, metaphysics is all we should care about? That is, that – since metaphysics studies the fundamental nature of the whole of reality by employing peculiar conceptual instruments that are independent of, and more fundamental than, those used in science – it can be expected to tell us how things are actually like independently of science? Hardly so. As pointed out earlier when considering the role of inference to the best explanation in science and metaphysics and the conjectural nature of both scientific and metaphysical claims, both disciplines start with questions about reality that arise from experience and are subsequently elaborated and assessed at the conceptual level. However, the conjectures and hypotheses that (may) constitute the outcome of this process need to be put to the test of reality itself, otherwise they remain pure hypotheses in the sense introduced above. And, as we tried to argue – in this agreeing with empiricist critics of metaphysics – while such confrontation with experience *is an integral part of science*, this is not and cannot be the case for metaphysics, exactly because of its nature and the nature of the concepts and categories it employs. Indeed, once one decides to insist that metaphysics is *essentially* defined as a discipline that proceeds entirely *a priori*, which appears inevitable if one is to avoid eliminativism, one cannot underestimate this point. Thus, it seems fair to say that, in opening to the test of experience, metaphysics *ipso facto* becomes (or at least gives way to) something else, i.e., science. It is a consequence of this that metaphysics, if it is to truly contribute to our knowledge of the world, cannot be independent of science.

On the other hand, we also argued, the foregoing does *not* mean that one is justified in demanding that metaphysics be made dependent on science. For, it is *exactly because* its mode of inquiry is radically different from that of science and it uses *sui generis* concepts and categories that metaphysics can be regarded as doing something valuable from a scientific perspective. In other words, the fact that metaphysicians employ peculiar concepts and categories and proceed *a priori* in their inquiry does represent a crucial difference between science and metaphysics. But, far from being a cause for condemnation and rejection, it should be regarded as the reason why metaphysics should not be reduced or set aside altogether by scientifically-minded philosophers (unless, we have already conceded, they are instrumentalists about science) and should instead be used to interpret science.[28]

But how exactly can metaphysics be at the same time non-self-sufficient (because it proceeds almost entirely *a priori* and can only become empirically relevant indirectly, by 'getting in contact' with science, as it were) and autonomous (because it inquires into a domain which is more fundamental than those studied by the special sciences and, in virtue of this, can contribute to the interpretation and understanding of the latter)? Should there not be some form of dependence between science and metaphysics? How exactly is metaphysics to be characterised once the Quinean picture of it is set aside, or at least given a secondary role? To provide an answer to these questions, and thus also a positive characterisation of the appropriate science-informed conception of metaphysics, we now need to get back to something that we only hinted at so far (when mentioning the ideas about possibility vs. actuality and ontological priority and dependence put forward by Lowe and Schaffer, respectively). This will be done in the next chapter. The basic ideas, however, should be already clear by now, and can therefore be expressed succinctly. On the one hand, there is no dependence relation either way between science and metaphysics because they are rather complementary activities, each one completing and supporting the other. On the other hand, there is no contradiction in the proposed characterisation of metaphysics because there really are two levels of metaphysics involved here. At a first level, one has metaphysics in the strict sense: it is an autonomous *a priori* activity which is independent of science but also, because of that, mostly lacking real informative content with respect to reality. At a second level, one has metaphysics in a looser sense, which really corresponds to the activity of applying metaphysics (in the strict sense) for the interpretation of our best current science – so also selecting among the available metaphysical hypotheses.

3. Conclusions: need for a compatibilist naturalism

Let us take stock. So far, we have seen that:

1) Metaphysics and science often explore the same parts of reality, and both employ abductive methods for evaluating competing explanatory hypotheses;
2) Metaphysics is *a priori* while science is based on observation and experiment;
3) Metaphysics seeks the most fundamental and general truths – those referring to which one can ground the interpretation of scientific claims;
4) Hence, metaphysics cannot be read off from science – reductionism about metaphysics is ruled out;
5) The tools of logical/conceptual analysis cannot lead metaphysics too far beyond the individuation of possible ways things could be – metaphysics isolated from science is too abstract;
6) Metaphysics can be 'fleshed out' via a consideration of our best current science – metaphysics can and must be science-based;

7) On the other hand, science is not 'understood' but merely usable as long as it is not interpreted, but the interpretation of theories does not come 'from within' – eliminativism about metaphysics is to be ruled out if instrumentalism about science (broadly understood) is.[29]

(5) and (6) convey the idea that, if there is something to the empiricist tradition and to its scepticism towards metaphysics, it is that it is very sensible to claim that all knowledge about the (material) world cannot but come from experience. If this is the case, however, it must be acknowledged explicitly that an *a priori* study of what is possibly the case can lead us to reach definite conclusions only in a very limited amount of cases. Namely, in those cases in which we can use logic to spot contradictions we did not notice beforehand.[30] In most cases, however, metaphysics does not tell us anything precise, and what it says about the possible must be evaluated and compared on the basis of our best source of information about what is actual, namely, science. In this sense, metaphysics as a study of an abstract domain of possible ways things could be like can be 'substantiated', as it were, only by making reference to science. At the same time (this is the other side of the coin), science requires metaphysics to be interpreted and, thus, properly understood – for a scientific theory does not and cannot come equipped with its own interpretation (this, notice, is not tantamount to saying that all scientific theories equally require an interpretation; there is probably a difference in this respect between, say, fundamental physics and molecular biology); and this means, to repeat, that metaphysics should appear necessary to all those who do not share a broadly understood instrumentalist approach to science itself. Indeed, if one is not interested in knowing what exactly we should think the world to be like if our best scientific theories are true, *a fortiori* one will not be interested in knowing what we should think the world to be like at the level of properties, objects, or persistence. For everyone else, however, taking science seriously appears to entail that metaphysics too should be taken seriously. (This is what (7) above says.)

It thus seems correct to claim that whether or not one should acknowledge the role of (proper) metaphysics alongside that of science is just a matter of personal preference (Chakravartty [2007]), that is, of one's personal stance in the sense of Van Fraassen [2002] – and this, notice, also in the sense that scientific realists need not engage in metaphysics: their position is not contradictory or anything like that if they don't.

Also it would seem that once what counts as proper metaphysics (at least from the perspective of the scientifically-oriented philosopher) has been identified, there should not be degrees of acceptance but only the fundamental yes-or-no attitude towards metaphysics. More precisely, the only sensible alternative seems to be the following: one could only do metaphysics as long as it can be immediately applied for the interpretation of science or, alternatively, develop metaphysics independently and then seek an application of parts of it. This does seem a choice that, again, can only be made on the basis of subjective preference. While it could be argued that the former approach is safer, the latter seems more in harmony with the view of metaphysics as an independent enterprise that can and should subsequently be put in relation with science. Indeed, in the same way in which, say, mathematicians study their field independently of possible applications in the empirical sciences, but mathematics is important for science to the extent that it can be so applied, so metaphysics can be regarded as a tool that is best developed on its own, independently of its practical uses, without this meaning that the naturalistic methodology and its fundamental presuppositions are *ipso facto* abandoned. It is believed here that this attitude is preferable, but a choice in this sense does not have to be presupposed for our purposes in the rest of the book.

Getting back to the discussion of naturalism as the legitimate request for continuity between science and metaphysics, we can, thus, now formulate the following definition:

> NATURALISM: Naturalism about metaphysics is best understood as the view that metaphysics should preserve its autonomy (we have suggested, as the *a priori* inquiry into a possibility space primarily characterised by dependence relations), but be studied in parallel with science, being put to the test of empirical evidence while at the same time defining the tools for the interpretation of science itself.

As already explained, there is no internal contradiction in this definition because there are two levels at which metaphysical inquiry is carried out, and thus two senses of metaphysics involved here. The first level is that of what counts as metaphysics in the strict sense. There, metaphysics is indeed an autonomous form of *a priori* inquiry. But at a second stage, the results of that inquiry can and should be put in relation to science. And there metaphysics turns out to be dependent, but it also stops being metaphysics in the strict sense of the term. Naturalised metaphysics, in the sense being defended here, then, truly is non-naturalised (in the traditional sense) metaphysics systematically confronted with empirical data.

Understood as suggested, naturalism does not entail the sort of eliminativist stance that seems to underpin both classical empiricism and the kind of naturalism currently exemplified, for instance, by Ritchie's deflationary attitude. Nor does it justify the rather restrictive conception of metaphysics endorsed by Ladyman and Ross. Rather, it invites one to opt for a form of *compatibilism* between science and metaphysics, intended as a general stance that takes both disciplines seriously and regards them as completing each other in essential respects. In other words, what one may call *compatibilist naturalism* amounts to a 'constructive' approach to metaphysics that gives up the idea that to naturalise it means to make it as similar as possible to science, and insists instead that metaphysics has to 'be in close contact', as it were, with the sciences while preserving its own autonomy and independence. This, of course, means to satisfy the 'continuity' element of naturalism without reading it as demanding an extension of the methods of science to metaphysics. As a matter of fact, even the idea of metaphysics as first philosophy might be preserved in this context – so perhaps reviving certain aspects of Aristotle's philosophy that deserve further attention if only from a methodological perspective.

Of course, it may be objected at this point that compatibilist naturalism is not naturalism enough, as the very definition of naturalism requires one to search for knowledge exclusively through science. This objection cannot be disposed of if not by pointing out that it rests upon a definition that one is free to question; and by adding that, since the very demarcation between science and metaphysics, as we have seen in this chapter, is not as sharp as is often implicitly assumed and, in fact, there are many points of contact and continuity between the two, it is indeed advisable to endorse a more flexible viewpoint.

What is one to gain if one accepts compatibilist naturalism? The thought is that if metaphysics is understood as a self-standing *sui generis* inquiry into the nature of reality that can, and should, develop autonomously but *in parallel with* science, then it becomes sensible to believe that philosophy and science can support each other and *converge towards deeper and more general hypotheses about, and descriptions of, the world*. This process is one which results from the systematic interplay of two dynamics:

a) One in which metaphysics as the *a priori* analysis of possible ways of categorising reality (although strictly speaking independent of science) is tested against the background of our most well-established scientific theories, so coming to be 'fleshed out', as it were;

b) One in which science (although strictly speaking independent of metaphysics) finds in the more encompassing (hence, in a sense, more ambitious) hypotheses and concepts of metaphysics a ground for both interpretation and further development.[31]

At this point, we need to articulate the proposed view as much as possible, so as to avoid the risk that it remains vaguely defined and, in the end, inapplicable. In particular, it is necessary, especially in view of the evident gap in the literature, to try to see how the proposed approach translates in practice; and, also, to define precise criteria of theory-evaluation and theory-choice, not only in metaphysics but when it comes to 'putting science and metaphysics together'. In other words, a methodology for naturalised metaphysics is needed. Especially so for the present work, which aims not only to present a broad meta-metaphysical perspective but also to apply it to specific case studies. In the next chapter, we will attempt to accomplish this task. The subsequent chapters will then embark in the enterprise of applying the tools of ('moderately' naturalist) metaphysics to specific questions at the boundary between science and philosophy.

Notes

1. It is perhaps worth emphasising that in the 20th century, science became once again closely connected to technology, so coming to have a practical function similar to the one it had for the ancient Egyptians or Babylonians. It goes without saying, though, that contemporary science is not regarded (exceptions, mostly among academics, notwithstanding) as only having instrumental value: rather, science is socially praised exactly as the fundamental source of knowledge about reality.
2. Essentially aiming to unify all knowledge on the basis of a standard scientific language, and to reduce all meaningful statements to sentences about elementary sense-data.
3. Bergmann was a member of the Vienna Circle, but he employed the general methodological principles set out for philosophical inquiry by the neopositivists with a view to defining a philosophical system that departed from basic neopositivist tenets in important respects, first and foremost in its endeavour to put forward clearly non-empirical theses.
4. As illustrated by Van Fraassen [2002; appendix B], what 'empiricism' and 'empiricist' exactly mean is far from obvious. For present purposes, suffice it to identify empiricism with the view that experience is the fundamental source of all knowledge.
5. Similar ideas are defended by Maddy [2007], who defines and defends what she calls 'second philosophy', a way of doing philosophy that she regards as a radical and austere form of naturalism. Essentially, second philosophy ignores the big philosophical questions and systems, and recommends instead the use of "what we typically describe with our rough and ready term 'scientific method'...without any definitive way of characterizing what that term entails [!]" [Ib.; 2].

6. Ladyman and Ross state that a lot of contemporary metaphysics that presents itself as taking science seriously can at best be regarded as 'philosophy of A-level chemistry' [Ib.; 24].

7. One may even suggest that their extension of structuralism to sciences other than physics is *ad hoc* – that is, put forward exclusively in order to lend support to the principle of naturalistic closure. Indeed, in discussing the objection that ontic structural realism might be right for physics but not for the rest of science, Ladyman and Ross explicitly claim that "this conclusion must be avoided by the naturalist as being inconsistent with the [principle of naturalistic closure]" [2007; 157].

8. Of course, an independent definition of what counts as metaphysics is still needed. If it were given in terms of explanatory unification, there would be no way to distinguish between unifying hypotheses that are metaphysical and unifying hypotheses that come from a specific empirical science, and one would be forced to give up the very distinction that the whole discussion rests on.

9. Another view worth mentioning at least in passing is Ney's [2012] 'neo-positivist metaphysics'. Resisting Ladyman and Ross's claims, Ney attempts to preserve a (limited) role for 'rationalist, armchair methods' by suggesting that, although there may not be objective grounds for adjudicating among different formulations of the same bit of physical theorising, metaphysics can help individuate those elements preserved in every formulation of fundamental physics and that appear indispensable based on the scientific community implicit or explicit attitude. Examples include determinism, laws of nature and probabilities in Everettian quantum mechanics (with respect to which, says Ney, a priori analysis is required in addition to the indications coming from the relevant theories). This is interesting, but it remains unclear (i) to what extent – granting that talk of indispensability is allowed at all – the work of metaphysicians is, on this construal, more than a merely mechanical drawing of consequences from facts about scientific theories; and (ii) whether the work done deserves the label 'metaphysics' or, to the contrary, shows that the very general hypotheses one is dealing with, contrary to what Ney (like the early Schlick) thinks, are scientific and not philosophical after all.

10. However, some philosophers do insist that metaphysics should be naturalistic in the sense that it shouldn't be in principle unable to have observable consequences. See, for instance, Maclaurin and Dyke [2012]. For a reply, usefully pointing out a striking connection with the neopositivist views on verifiability and their failure, see McLeod and Parsons [2012]. Basically, as was pointed out against Ayer's criterion of 'factualness', every theory can be made to have observable consequences by aptly adding auxiliaries to it. For a rejoinder, see Maclaurin and Dyke [2013].

11. If only for completeness' sake, it is worth mentioning here the project of naturalisation of metaphysics proposed by Goldman [2007]. Basically, Goldman suggests that the empirical study of the human mind should be a fundamental part of metaphysical investigation. Via practical examples, Goldman shows that many presuppositions that we may regard as obvious, for example, concerning the individuation of events, actually depend on the specific way in which our mind works with respect to the environment. Thus, he concludes, rather than doing traditional philosophy on the basis of

these presuppositions we should instead make the latter our primary object of inquiry. While it is certainly right that we should problematise and study via empirical methods everything that we can, including first and foremost our conceptual structures, the sort of naturalisation proposed by Goldman hardly qualifies as a candidate for understanding the connection between science and metaphysics in general, which lies at a higher methodological and theoretical level.

12. It should be clear that by 'science' I mean here, and shall mean in what follows, natural, empirical science.

13. It is evident, then, that the Greek word 'meta', is not to be intended here only as the English 'after', but as 'beyond', taken in the sense of 'more fundamental'.

14. Ontology can be either 'general' or (to use a term introduced by Husserl) 'regional', that is, applied to a limited domain of things. Moreover, there is a relevant distinction to be drawn between 'formal' and 'material' ontology, the former studying the abstract features of certain classifications of entities, and the latter specific classifications of entities, respectively. Here, we are, and will be, concerned exclusively with general material ontology.

15. The formulation used here is motivated by the fact that, strictly speaking, the thesis of materialism and that of physicalism are not (necessarily) identical (see Stoljar [2010]).

16. The debate on this is of course complex, and involves discussion of at least three sets of issues: (i) Whether a Theory of Everything is possible, or it is to be ruled out in principle – for example on the basis of Gödel's incompleteness results; (ii) Whether we would be in a position to recognise a (the?) Theory of Everything as such, were we to happen to formulate it; (iii) Whether acceptance of the possibility of a (the) Theory of Everything doesn't presuppose precise solutions to important philosophical problems such as those surrounding physicalism and inter-theory reduction. All this, at any rate, is immaterial with respect to the more general point being made here.

17. Namely, that it has not discovered the Theory of Everything yet.

18. Consider, for instance, the question whether Goldbach's conjecture – that every even integer greater than 2 can be expressed as the sum of two primes – is true or false.

19. And thus does not deliver not only what the neopositivists were looking for, but also what contemporary defenders of neopositivist naturalised metaphysics are looking for (see the references to the work of Ney and Maclaurin and Dyke in notes 9 and 10 above).

20. I am assuming here that the idea that the Earth is spherical was first formulated (as common wisdom has it, by Pythagoras) entirely independently of observational facts that were at least indirectly relevant for it (e.g., shadows of objects, dynamics of eclipses, positions of constellations), even though the latter started to be known relatively early, way before direct testability (for instance, observation by astronauts) became available. Indeed, in this sense, the spherical Earth example appears to be paradigmatic of the heuristic role attributed to metaphysics by, for instance, Popper.

21. This is interestingly analogous to the debate concerning whether string theories are contingently or in principle incapable of providing testable predictions, and what would entitle one to deem them 'metaphysical'. On this,

too, the debate is still open, and uncontroversial criteria of assessment still lacking.

22. This is not exactly right, as there are ways for bundle theorists to avoid a commitment to the necessary truth of the Identity of the Indiscernibles (see Rodriguez-Pereyra [2004]). At any rate, this does not affect the point being made here.

23. With this, two things should become clear. First, that allegedly metaphysical hypotheses that turn out to be directly testable, according to the present proposal, *ipso facto* cease to be metaphysical. Secondly, that, as a direct consequence of this, the conception of metaphysics as the mere search for the most general claims that can be extracted from science (which, as we have seen, is defended for instance by Ney [2012]) is not satisfactory because it diminishes the potential fruitfulness of metaphysics. Not surprisingly, an enemy of metaphysics such as Van Fraassen is also happy to acknowledge that metaphysics aims at "excavating the ontic commitments buried deep inside the empirical sciences" [2002; 197].

24. Quine says, "this is, essentially, the only way we can involve ourselves in ontological commitments: by our use of bound variables [...;] a theory is committed to those and only those entities to which the bound variables of the theory must be capable of referring in order that the affirmations made in the theory be true" [1951; 12–14].

25. Things are not so simple, for it can plausibly be argued that Quine 'only' added to Carnap's entirely pragmatic stance towards the choice of the linguistic framework to be used in relation to a certain domain (i) the label 'ontology' and (ii) the preference for a physicalist framework when it comes to inquiring into the deep structure of reality together with (iii) a rejection of a sharp distinction between science and philosophy via his critique of the analytic/synthetic distinction (leading to the demotion of scientific claims to the level of non-scientific ones). Based on this, one could argue that Carnap and Quine were not truly differentiated by their attitudes towards ontology. These issues of reconstruction, however, can be set aside here.

26. Also see Fine [2009] and [2012a].

27. Additionally, as correctly pointed out for instance by Maudlin [2007; essay 3, esp. 80–96], the Quinean idea of extracting ontological commitment directly from the domain of quantification necessary for the sentences of a given framework (once appropriately formalised) to be true presupposes a structure of objects with pure properties that can be questioned (a pure property being one that could be instantiated in a world in which only the item that has the property exists, the same holding for relations). Maudlin then goes on to actually question the picture and propose an alternative, based on fibre bundles, which he takes to receive much more support by contemporary physics.

28. Rather than to seek Truth on its own, of course. The empiricist critique to metaphysics is to be accepted wholeheartedly if metaphysics presents itself as capable of finding objective truths by itself (in any other way, at any rate, than by discovering logical inconsistencies).

29. That taking metaphysics to be truly relevant and not just merely a conceptual game requires scientific realism may seem obvious, at least in a naturalistic context, but is worth reminding nonetheless. In connection to this, it is

also worth noticing that actually using science to ground our metaphysical claims is more controversial than it may seem. Monton [2011], for instance, reminds us that – even if we are realists about specific theories, or physics as a whole – we cannot ignore the fact that current physical theories are literally false (idealisation and abstraction are always in play), often incompatible with one another, open to interpretation and, most importantly, the result of a historical process of continuous revision that entitles us to regard them, at best, as approximately true. Monton concludes that, if one does not want to wait for a better physics or restrict one's claims to the few parts of physics that are truly uncontroversial, one should acknowledge that metaphysical claims can, at most, be conditional claims ('If theory T is true, then...'), and in most cases present us with a number of possible alternatives. While Monton does not seem enthusiastic about this, it seems perfectly in agreement with the views expressed here. In relation to this, the fact that metaphysical claims are not presented in conditional form by no means indicates that they are not known to be dependent on certain assumptions being in fact correct. (The distinction just drawn between metaphysics in the strict and loose sense is relevant here.)

30. And, if they exist, at all, in cases in which whatever additional criteria (for instance, simplicity) we may wish to apply lead us to clearly prefer certain metaphysical hypotheses over others. But more on non-empirical criteria of theory-choice outside of science in the next chapter.

31. One may also interpret this as pointing to the fact that *a priori* and *a posteriori* are not, in fact, as sharply distinct as is normally believed, and are instead in some sort of mutual 'bootstrapping' relationship (see Tahko [2011]). We will not take sides on this issue here.

2
Naturalism

1. Possibility, grounding and constructive naturalism

So far, we have suggested that metaphysics must:

i) Proceed *a priori*;
ii) Employ peculiar concepts and categories;

but at the same time

iii) Be conceived of as in need of 'content', and capable of providing knowledge of the actual world only to the extent that it is functional to the interpretation of scientific theories.

To satisfy these requirements, we argued, metaphysics should be differentiated in important ways from the canonical understanding of it that more or less directly descends from Quine's reflections on what there is and on ontological commitment. In particular, we suggested in passing that:

a) Metaphysics is a study of a *sui generis* possibility space;
b) Metaphysics is primarily aimed to uncover the sort of priority and dependence relations – holding between (kinds of) entities and facts – that essentially characterise reality.

It is now time to say more about these two features.

1.1 Claims of possibility and kinds of modality

As for the first idea, having to do with possibility, as we mentioned it is in particular Lowe ([2001], [2011]) who recently distinguished in a forceful manner good and bad metaphysics along those lines. Lowe agrees that certain questions that are commonly regarded as metaphysical are completely idle ones, and that if Quine's characterisation of ontology were all there is to say about the nature of metaphysics, then metaphysics would indeed amount to little more than playing with a series of futile questions with no contact whatsoever with reality [2011; 101]. Having acknowledged this, though, Lowe also explicitly claims that

> metaphysics in fact compares *very badly* with empirical science according to this conception. However, ... the fault lies here not with metaphysics as such ... but only with those of its false friends who mistakenly seek to enhance its credit by assimilating its task to that of empirical science. [Ib.]

Clearly, Lowe's thought here is that it is wrong to attempt to naturalise metaphysics in the stronger sense of the term – that is, to try to satisfy the requirement of 'continuity' between science and metaphysics by reducing the objectives and methods of the latter to those of the former. As an alternative, Lowe proposes a view of metaphysics as the study not of how the world *is*, but rather of how it *could be*. Rejecting the project of naturalisation of metaphysics intended in the sense that the empirical methods typical of the sciences should be extended to philosophy, that is, Lowe recommends the pursuit of metaphysics as an independent *a priori* research aimed to

> envisage, in a very general way, what sorts of things there *could be* in the world, at its most fundamental level of organization or structure, and then develop arguments for or against the existence of things of this or that general sort. [Ib.; 104]

This view is endorsed here: metaphysics is a peculiar study of a possibility space prior to science, through which a complete grid of hypothetical options is defined that are then to be evaluated and critically compared. Of course, metaphysicians typically do undertake such an evaluation and critical comparison. They do so via non-empirical means. And it is probably this sort of arguments that Lowe has in mind in the above quotation. Here, however, the idea is that this process must be carried out, first and foremost, on the basis of the best available knowledge of what is *actual*, hence science. Lowe himself acknowledges that

> when both are conducted fruitfully, metaphysics and empirical science exist in a symbiotic relationship, in which each *complements* each other. ([Ib.; 102])

However, Lowe's leading slogan is that 'possibility precedes actuality', both conceptually and methodologically, and thus this complementarity does not detract from the overall precedence of metaphysics over science. Possibility, obviously enough, is in play also in scientific theorising, both in the sense that hypotheses are underdetermined by the evidence and in the sense that theories also account for what *could* happen or be the case. The sense, just illustrated, in which possibility is involved in metaphysical analysis appears however different and more fundamental.

Ladyman and Ross [2007], object that philosophers have often been wrong in deeming something possible or impossible, and have been shown to be wrong by the developments of science, which, as a matter of fact, always turned out to deserve priority.[1] This objection, however, is not compelling. First of all, it is true that philosophers have been often wrong in claiming that something is possibly/necessarily (not) the case, as the contrary turned out to hold as science progressed. But starting from the view – endorsed here – that the aim of metaphysics is to identify the range of what *might* be true, and only in limiting cases[2] can the metaphysician point to (what he or she perceives as) *the* actually true hypothesis, a few rejoinders to Ladyman and Ross remain available. First, the examples offered by Ladyman and Ross might be interpreted as only showing that some philosophers were not doing proper metaphysics and/or were misunderstanding the nature and capacity of proper metaphysics. Secondly, one may draw a distinction between one's *practical* and one's *philosophical* attitude towards a theory: the way in which one presents and discusses one's hypotheses does not necessarily mirror their kind and degree of commitment towards those hypotheses. Indeed, practising scientists too display a realist attitude towards their theories when it comes to presenting and/ or defending them – and not only at the didactic or popular science level; yet, when pressed, many (if not most) of them turn out to be essentially instrumentalists. Third, one may point out that in no way do we take the philosophical attitudes of individual scientists to necessarily mirror essential characteristics of science itself rather than just contingent, socially determined preferences and beliefs; and thus, even if all scientists, past and present, turned out to have been and be convinced realists about their theories, we would not take this to make science less respectable in light of possible and actual episodes of falsification contradicting such a realist attitude. If this is correct about science and scientists, why should one not be allowed to take an analogous stance towards metaphysics and metaphysicians? Lastly, even if one takes the mistaken philosophers that Ladyman and Ross have in mind here to be truly paradigmatic of metaphysics, it can, in any case, be objected to Ladyman and Ross that they define criteria of acceptability for metaphysics which are too strict according to their own standards. For, if it is true that metaphysics should be pursued only to the extent that it is connected to science and its methodology, why consider fallibility (commonly regarded as a, if not the, distinctive feature of science) a fatal problem and not a strength for metaphysics? Notice, in this connection, that people like, for instance, Kant, in spite of the falsity of their metaphysical claims (in Kant's case, for instance, concerning the real geometry of the physical universe), were definitely in contact with the best science *of their time*. What else should be required from the naturalistic perspective?[3] As a matter of fact, as we will see in more detail later (Chapter 3), Ladyman and Ross' own metaphysical views are arrived at via a prior critical comparison of metaphysical alternatives. Given this, one might go as far as to say that it would be self-contradictory for them to reject the depiction of metaphysics as an autonomous discipline dealing with possibilities being offered here. At most, then, what follows from what Ladyman and Ross say about the space of alternatives into which metaphysics moves is that

i) We become aware of the structure of such possibility space and of the different items that populate it in a gradual fashion;
ii) We do so in a way that is necessarily affected by our experience and knowledge of the actual world.

But at no point is the (naturalised) metaphysician forced to deny this.

On a related note, the very idea that there is a peculiar possibility space for metaphysics to study might be put into doubt on the basis that, in the end, it is *only* science that tells us what is possible and what is impossible. This might be what Ladyman and Ross had in mind in their book. At any rate, it certainly is the complaint expressed by, for instance, Callender [2011]. In the course of an interesting examination of the relationship between philosophy of science and metaphysics, Callender protests that a clear definition of this supposed peculiarly metaphysical possibility space is lacking. This he takes to mean, primarily, that metaphysical *modality* has not been made plainly distinct from physical/causal/nomological/natural[4] modality. This, in turn, suggests that Callender's idea is – at least implicitly – that those defending the autonomy of metaphysics and its irreducibility to science should take pains to unambiguously identify metaphysical modality and show it to be itself irreducible. But Callender is sceptical about this:

there is no interesting species of metaphysical modality that is largely immune to science. Our modal intuitions are historically conditioned and possibly unreliable and inconsistent. The only way to weed out the good from the bad is to see what results from a comprehensive theory that seriously attempts to model some or all of the actual world. If the intuitions are merely 'stray' ones, then they are not ones to heed in ontology. *In metaphysics we should take possibilities and necessities only as seriously as the theories that generate them.* [2011; 44]

Now, does the claimed dependence of metaphysical modality on physical modality suffice for rejecting metaphysics even in its naturalistic flavour (regardless of whether or not this was Callender's intention)? My goal in the rest of this sub-section is to answer this question in the negative, by denying that the (limited) autonomy of metaphysics requires the irreducibility of metaphysical modality. Indeed, Callender is certainly right that

metaphysical modality is murky [...and] at the juncture of many disputes in philosophy of language, mind and logic [...]; and]there is nothing to be found in Kripke's examples that would warrant thinking of metaphysical possibility as something immune to actual science. [Ib.]

Even stronger theses have been put forward, for instance by Leeds [2007], who makes the perfectly general claim that physically possible worlds are all the possible worlds there are. (In particular, Leeds argues that the "core metaphysical necessities are those X such that X is (physically) necessary, and such that we cannot conceive, for an *appropriate* true A, that (A & (physically) possibly ¬X)" [Ib.; 471] – appropriateness meaning here that the content of the relevant modal claim is correctly grasped, A=X being the limiting case). But even if one were right in deflating in this fashion the autonomy and priority of metaphysical necessity in favour of physical necessity,[5] our earlier claims about the function of *a priori* analysis would remain valid. For, all is needed for those claims to hold is that meaningful questions can be asked (for instance, 'Does free will exist?' or 'Is space a substance or a network of relations?') that science leaves necessarily underdetermined, and that can receive instead tentative answers through alternative – and, we have contended, equally rational – methods. And this requires only reality to be described on the basis of a peculiar *vocabulary* and employing peculiar *categories* that are different from those of science and not definable in terms of the latter. Since linguistic/conceptual irreducibility by no means requires, or even just suggests, the existence of distinct modal domains, complaints about the vagueness of the notion of metaphysical modality – if taken to point to more than the need to use science as a guide in 'fleshing out' our *a priori* thinking – are in fact misplaced.[6]

The foregoing might have seemed too quick. Let us then look at the issue in a bit more detail. The key issue is whether and how metaphysical modality relates to *conceivability*. Indeed, what most (if not all) supporters of the view that metaphysical possibility truly is nothing more than physical possibility seem to think is exactly that what we find conceivable is determined by experience only, and that metaphysical possibility is to be defined in terms of conceivability, hence metaphysical possibility is determined by experience only (recall the above references to Callender and Leeds). And conceivability has in fact very often been referred to in defining metaphysical possibility. Today, it is still – or, perhaps, again – the most authoritative candidate for analysing metaphysical modality.[7] In a seminal paper, Yablo [1993] first examines various senses of conceivability (e.g., as believability) and explains why they are insufficient for accounting for metaphysical modality. Next, he proposes that

i) Something P is conceivable for a subject x if x can imagine a world that s/he takes to verify P;

and, similarly,

ii) Subject x will find P inconceivable if x cannot imagine a world that s/he doesn't take to fail to verify P.[8]

But, in order to determine what we can imagine we cannot but use what we know about what is actually the case and 'manipulate' it, as it were, in our minds. And this seems to suggest reductionism, because – setting the logical/conceptual element aside – all that counts seems to be the knowledge of things that we obtain through empirical methods, i.e., of physical modality. A related difficulty concerns how to truly differentiate the proposed account of metaphysics from Carnap's internal questions and from Quine's views of ontological commitment as dependent on the specific theoretical framework and its appropriate formal regimentation. For, if what is metaphysically possible is what we find conceivable from within the conceptual scheme(s) defined by our best knowledge of the actual world, then metaphysical questions reduce to internal questions, and metaphysical claims to claims to be evaluated on the basis of the structure of those conceptual scheme(s).

Against the foregoing considerations supposedly threatening the autonomy of metaphysics, the following considerations can be brought to bear. First, that what we can or cannot conceive is determined on the basis of empirical knowledge 'manipulated' on the basis of logical/conceptual constraints does *not* mean that the resulting scenarios have to be describable entirely in terms of the vocabulary of the relevant empirical sciences. To the contrary, far from just structuring, say, sentences of current physical theories via logical connectives, when we conceive a possible world – even if we do it entirely on the basis of what we know about the actual world through science – we can (and often do) do this by introducing new concepts and categories. For instance, scientists take for granted and use the concepts of a property or of qualitative sameness. In view of this, we can, on the basis of what they tell us, ask whether the properties they refer to are tropes or universals, and whether the similarity facts they describe are primitive or reducible to numerical identity of instantiated properties. And this affects our determination of what worlds count as possible. In a word, even if, at root, all necessity were natural necessity, this would not lead to the elimination or reduction of the metaphysical vocabulary and the corresponding conceptual categories to those employed in the sciences.[9] And, obviously enough, allowing for meaningful descriptions of possible scenarios that transcend those provided entirely in the vocabulary of the sciences does not make those descriptions useless: 'metaphysical facts' might (and are likely to) just be special facts about the actual world normally studied by the sciences, which are not subject to direct empirical inquiry and yet potentially relevant for our understanding of things.

The foregoing suffices for neutralising worries such as Callender's: to repeat, the respectability and (limited) autonomy of metaphysics does not require the irreducibility of metaphysical modality. However, supporters of metaphysical *modality* – not just metaphysics – could also aim higher. In connection to this, the issue of monism versus pluralism about modality becomes relevant. Monists believe that there is only one fundamental kind of modality – usually identified with either logical or metaphysical modality. Indeed, one may think that monism (of course, with priority given to metaphysical modality) is necessary for those who want to avoid reductionism about metaphysics, and so is a definition of physical necessity as what results from implementing a 'restriction' on the broader domain of metaphysical necessity. While this is an open possibility that we don't need to rule out, positive arguments can be (and have been) given in favour of pluralism. Fine [2002], for instance, contends that neither physical nor normative necessity and possibility can plausibly be regarded as restricted or relativised forms of metaphysical necessity/possibility.[10] And he also provides arguments against forms of monism reducing all modalities to natural necessity.[11] Clearly, on this construal, it becomes possible for the constructive naturalist to accept the dependence (via conceivability) of metaphysical modality on physical modality. For, he or she can at the same time insist that *both* correspond to objective facts 'out there', exploiting putative counterexamples to the reducibility of one type of modality (in this case, metaphysical modality) to another (in this case, physical modality).[12]

It thus seems that naturalistic metaphysicians have at least two ways – one more 'modest', the other more 'ambitious' – to circumvent problems having to do with the supposed parasitic nature of metaphysical possibility and necessity claims on physical possibility and necessity claims. Regardless of whether one opts for pluralism about modality and consequently emphasises the irreducibility of metaphysical modality, or 'just' accepts the reducibility of metaphysical to physical modality while resisting the parallel reduction of metaphysical to physical conceiving and theorising,[13] the important point is that sceptical arguments against the autonomy of metaphysical modality do not prove constructive naturalism about metaphysics unworkable.[14]

Before closing this sub-section, another important point related to the status of metaphysical modality needs to be discussed: the role played by the notion of essence in the definition and determination of metaphysical possibility and necessity.

1.1.1   Essences

In his definition of good metaphysics as the science of the possible, Lowe explicitly states that metaphysics is also the science of the essence of entities [2011; 99–100 and 105], that is, of the nature of things, of what things are and what being those things amounts to. Essences and essential properties are also invoked in other contexts, for example when defining, in a broadly Aristotelian–Lockean fashion, the sorts of substances that exist; or when trying to single out by empirical means the natural kinds that populate our world. This might mean that, indeed, the modal structure of the world really corresponds to the essences of things, but it also calls for some differentiations.

It does indeed seem right to think that metaphysical facts, i.e., those pertaining to the fundamental structure of reality, must have to do with essential features of that structure. How else could be the corresponding claims of possibility and necessity obtain their truth values? This will appear particularly compelling to those convinced by Fine's [1994] well-known arguments that essence is prior to necessity and, thus, it is not definable in modal terms. That is, if it is correct to think that what is essential cannot be reduced to what is necessary, but rather the converse, then essences should be given a central role in one's metaphysics. And this should lead the serious metaphysician to make room for essences in his or her metaphysical construal.[15] Consequently, essences should play a role in the context of (constructive) naturalism about metaphysics, too.

But, notice, all this only means that the metaphysician is required to be committed to the existence of essential features of *fundamental types of things*, i.e., of *general essences* of ontological categories (and, perhaps, of essential *relations* of dependence and priority in addition to those – more on this in a moment). Nothing must be said concerning supposedly essential properties of more specific kinds of things studied by special sciences; and even less concerning the *particular* essential properties of particular things. For instance, it seems perfectly possible to do metaphysics by only being committed, say, to the existence of essentially repeatable properties (universals), without saying anything about which empirical properties are exemplified by which kinds of objects; or to claim that it is in the essence of time to consist of relations between objects or between events, without saying anything about the nature of specific processes taking place in time and/or specific objects existing in time. In a word, the essences required for metaphysics to be an informative inquiry into the possible correspond to the genuinely metaphysical properties of very general families of things, not 'just' to empirical properties discovered to be essential to things by a posteriori means. In this sense, although they are commonly regarded as paradigmatic of what metaphysical necessity consists of, and of the amount of independence and autonomy that must be granted to metaphysics, the traditional Kripkean examples of *a posteriori* necessities are in fact not particularly important, and possibly also misleading – at least with respect to our present purposes. The fundamental aim of metaphysics is, to repeat, not to determine the contents of specific *a posteriori* necessary statements but rather to discover essential features of the general structure of reality.[16]

Notice that the foregoing does not imply that one should endorse the essentialist view of modality – according to which metaphysical modality is the most general type of modality because it reduces to the essences of things – that seems to be defended, among others, by Lowe [2001] and Shalkowski [1997]. Such a view requires specific assumptions about the nature of essence and our knowledge of it which we explicitly refrain from making here.

This discussion of essences completes our answer to Callender's worries, reported at the beginning of this section. (In light of our previous discussions, it is hardly worth adding to this that this commitment to 'general essences' does not require one to think that such things can only be inquired into as objective features of reality if the complete autonomy and independence of metaphysical modality is also assumed.)

1.2   Existence and grounding

Let us now consider the second key element in the proposed conception of naturalised metaphysics: that is, the idea that metaphysics should be regarded as (primarily) a study of grounding relations. Previously, we have argued that the Quinean conception of ontology and metaphysics is indeed, at least to some extent, problematic as detractors of metaphysics suggest. But we also claimed that such a conception is not compelling, and in fact deceptive. Following Schaffer [2009], we agreed with the claim that metaphysics is best intended as a study of *grounding relations*, that is, of relations of dependence and priority between (categories of) things – relations that are not of a causal nature and, consequently, cannot be reduced to anything studied by the natural sciences. (These relations might, however, have to do with essences intended as suggested in the previous sub-section.) According to Schaffer, when we start asking questions about what is more fundamental than what, and what is reducible to/dependent on/analysable in terms of what, we can see that truly interesting questions emerge.[17] And that these are neither (at least not always) reducible to scientific questions, nor to be eliminated as remote from the empirical evidence in Van Fraassen's sense discussed in the previous chapter. Consider, for example, the issue, already mentioned earlier, concerning the nature of identity and individuality in quantum mechanics (an issue to that will be dealt with in detail in Chapter 3). Does the evidence coming from the quantum domain tell us what we should believe about objects and their identity conditions, or is a general theory of the relationship between identity facts and qualitative facts a pre-requisite for a proper understanding of the empirical data as they are described by our best current theory? At least if the input coming from physical science is to be interpreted and, hence, further elaborated to be understood, the second alternative appears more plausible. And, notice, this amounts to saying that an independent philosophical assessment of the priority or dependence relations holding between qualitative facts and identity facts – or, better, a prior assessment of the *various* ways in which facts of these two types *might* be mutually related – is required. This, we want to claim here, can be generalised:[18] although perhaps not all metaphysical questions that can be fruitfully asked while putting them in connection with current science need to be about priority and dependence, these sorts of questions are nevertheless fundamental for a proper characterisation and assessment of metaphysics from a scientifically-informed perspective. To repeat, the latter should not and cannot be limited to the canonical characterisation of ontology – which, Schaffer argues, in fact *presupposes* the notion of grounding as fundamental: in identifying the best scientific theory or the canonical logic, but also the apt linguistic translations, the right domain and, indeed, the ontic commitments implicit in our best theoretical constructions.[19]

On the other hand, the Quinean approach need not be dispensed with altogether: the suggestion is simply that 'what (kinds of) things are there?' is not the fundamental question, especially not when considered on its own, independently of questions concerning the mutual relationships between various types of entities and facts.[20] Among other things, it follows from this that the way in which metaphysics should be done, and its outcomes evaluated, changes: for instance, a more permissive catalogue of entities might be preferred to a more economic one (something that the Quinean would regard as wrong) if the structure of dependence relations that it gives rise to is better (for instance, simpler, more explanatory etc.).

At this point, a possible difficulty must be taken into account. Several authors objected to the very notion of ground is confused and/or incoherent or of no real use (see, for instance, the discussions in Hofweber [2009] or Fine [2010]). If this were the case, obviously enough, the alternative proposed by Schaffer could hardly be a promising starting point for constructive naturalists. To be sure, the concept of grounding does have to be clarified, and there are many aspects of it that call for further philosophical work. A quick way out consists in claiming that this should not be regarded as a surprise, as the notion of ground has become the subject of careful systematic discussion only recently. Whether or not this is considered too easy, even independently of future developments in this area of philosophy it can be plausibly argued that the critiques mentioned a moment ago are far from conclusive. The charge of confusion, for instance, can be neutralised by either arguing (i) that the notion of ground really is basic, and identical with ontological dependence, perhaps restricted to actual facts in the world,[21] or (ii) the relation of grounding may not be fundamental but at the same time cannot be reduced to, or identified with, any other more familiar relation (such as that of supervenience, that of truth-making or that of causality) and is instead, best regarded as encompassing various philosophically significant notions (e.g., logical entailment, counterfactual connection) in a useful manner – for instance, because it highlights common features such as asymmetric connection or presupposition of a hierarchical structure.[22] As far as it is acknowledged that there is a metaphysical structure to reality, and that one or more relations are fundamental in that they determine such structure, it is not necessary here to choose between (i) and (ii), and we can consequently suspend our judgment and wait for further theoretical developments. As for the incoherence objection, that the notion of ground is logically incoherent (Fine [2010] offers the example of the fact that everything exists, which appears, by definition, to obtain (partly) in virtue of itself) can be resolved by noticing that classical logic is not completely untouchable and it might be both possible and useful to find – in Fine's words – 'some kind of reflective equilibrium' between the principles of classical logic and those it seems to conflict with (in this case, the principles of ground). A perhaps more worrying objection is that the notion of ground is (or may be) logically coherent but in any case belongs to an unappealing 'esoteric' metaphysical project aiming to answer certain questions by using a 'technical' vocabulary rather than ordinary terms accessible to all [Hofweber 2009; 266–267]. In this case, as well argued by Raven [2012], the answer is that a notion being technical, or 'esoteric' is definitely not a reason for not making use of it: for instance, mathematics is also an esoteric project (because it involves, e.g., a notion of set which is definitely not the naïve – and notoriously inconsistent – notion of a collection of entities), but we do not doubt that it is worth doing.

In light of all this, while of course the usefulness of the notion of ground can still legitimately be put into question, from now on we will assume that such a notion plays a crucial role in metaphysics. This follows directly from the assumptions that, as argued, (i) metaphysical questions are basically questions about the essences of things and their mutual relations in a structured world, and (ii) these relations are fundamental dependence and priority relations. As explained, here too, as in the case of possibility/necessity in connection with metaphysical modality, only a minimal commitment to the key notions and concepts is intended.[23]

More generally, then, there is reason to believe that the idea of 'possibility space' and the concept of metaphysical ground can constitute the basis for an interesting form of metaphysics, which can, in turn, constitute the foundation for a naturalistic approach to philosophy that is not unnecessarily eliminative or reductive. And this is exactly what the thing that we labelled constructive naturalism here is intended to be. Having said this, it is now time to see how constructive naturalism about metaphysics should be translated into philosophical practice. How is it, that is, that we should go about putting metaphysical hypotheses and scientific theories together? Can a precise methodology be defined, at least to some extent?

2. Experimental metaphysics and constructive naturalism

An interesting view of the metaphysical import of science, and in particular of physics, has been developed in the last 30 years or so, and is known as 'experimental metaphysics'. This concept was first introduced by Shimony [1980] and subsequently employed by other authors (for instance, Hellmann [1982], Jarrett [1989] and Redhead [1996; chapter 3]). Shimony explicitly defined it in the context of a discussion of quantum mechanics, and in particular of Bell's inequalities and the experimental confirmation of their violation, which is thus useful to look at (also in view of the fact that the evidence that we are about to illustrate will play a central role in later sections).

In his [1964], Bell revisited the well-known EPR paradox, presented by Einstein and two co-workers (Podolsky and Rosen) [1935] as the basis for an argument against the idea that quantum mechanics is complete (i.e., there is nothing that it does not represent of the physical systems it is concerned with). The original argument was based on an assumption of *local action*, or *locality*, formulated in harmony with special relativity. In Einstein's words, the assumption was that:

"It is characteristic of ... physical things that they are conceived of as being arranged in a space-time continuum. Further, it appears to be essential for this arrangement of the things introduced in physics that, at a specific time, these things claim an existence independent of one another, insofar as these things "lie in different parts of space". [... F]or the relative independence of spatially distant things A and B, this idea is characteristic: an external influence on A has no immediate effect on B; this is known as the principle of 'local action'.... The complete suspension of this basic principle would make impossible the idea of the existence of (quasi-)closed systems and, thereby, the establishment of empirically testable laws in the sense familiar to us". ([1948; 321–322], translation in Howard [1985; 187–188])

Einstein, Podolski and Rosen (EPR) considered a source emitting one at a time pairs of electrons in the singlet state of spin. These electrons do not possess determinate values of spin, but only correlated probabilities as regards measurement results. In particular, they have equal probabilities of being spin 'up' or spin 'down' along each spatial axis. The electrons are directed towards distinct measuring apparatuses, and one of them is measured first. The outcome of this measurement is a determinate value of spin. What is striking is that, once this measurement takes place, the spin component of the second electron (along the same axis as the spin measured on the first) is also determined, before a measurement takes place on it, even though the two electrons are at that point space-like separated (that is, at a distance that – if Einstein's principle of local action holds – seemingly rules out a direct causal connection). In particular, one of the two outcomes available to (and equally probable for) the second electron invariably occurs, namely, the opposite of that obtained in the measurement on the first electron. Einstein thought that this was enough to conclude that quantum mechanics is incomplete, as locality – he thought – cannot possibly be given up, and so there must be something about the particles, not described by the theory, which determines the evidence while in agreement with local action.

Bell, however, [1964] examined the data and the conceptual constructions in play more closely, and concluded that this is not the case. For, if it were, one should, in principle, be able to find 'hidden variables' (that is, additional physical factors the consideration of which makes the theory complete) enabling one to explain the evidence without having to admit of the existence of non-local causal connections. Relatively simple calculations (involving the violation of certain equations now known as 'Bell inequalities') show, however, that any such hidden variable theory is bound to be non-local anyway. Hence, it seems that quantum mechanics entails the rejection of Einstein's metaphysical presupposition of locality.[24]

According to Shimony [1980], a general pattern can be individuated in cases such as the one just described. It has the form E&H⇒P, where E is an accepted theory used to describe the relevant experimental setup (here, quantum mechanics as it is employed to perform actual tests of Bell's inequalities), H represents a general metaphysical hypothesis (in this case, locality), and P signifies a certain empirical prediction (in the above example, that the Bell inequalities hold). If P is disconfirmed, and E is kept fixed, says Shimony, by *modus tollens* we get a rejection or modification of H, so bringing experiment to bear upon a metaphysical thesis. In Shimony's words, it seems that:

> Bell has provided us with the means for treating certain metaphysical hypotheses with the same level of control that has been achieved for typical physical hypotheses. [1981; 572–573]

This claim is remarkable, as it asserts that we are in a position to learn metaphysical lessons on the basis of experimental data, logic and suggestions taken from our best current theories. Perhaps, then, the constructive naturalist really does nothing more than propose a generalised use of the methods of experimental metaphysics. That is, that experimental evidence be systematically brought to bear on metaphysical hypotheses via deduction of testable predictions and the application of *modus tollens*. Indeed, given what we have said so far, it does look like the dynamics identified by Shimony is the right way of making metaphysical conjectures and actual science interact. And this even goes some way towards satisfying the requests of the more radical empiricists.

However, while all this seems plausible, and to a large extent correct, there is more to say, as a description of experimental metaphysics by no means suffices for providing an exhaustive characterisation of constructive naturalism. There are three points to be made:

1. First of all, even in the specific case of quantum mechanics and relativistic locality, there is much more going on than the mere application of *modus tollens*. One possible reaction to the evidence described above is to claim that a 'peaceful coexistence' between relativity and quantum mechanics (that is, between locality and quantum correlations, i.e., E and H above) should be sought. Some authors believe that it can, indeed, be found. Jarrett [1984] argues that the failure of the Bell inequalities can, in fact, be connected to the violation of either one of two different conditions: either a locality condition – different from Einstein's local action (and thus better defined as *parameter independence*, for reasons that will become apparent shortly) – which states that the outcome of the measurement at one end of the experimental setup is statistically independent of what is measured at the other (and of all the factors determining the exact nature of the other wing of the measuring device); or a completeness condition (or, *outcome independence*), according to which the probability of the joint outcomes, given the components measured and all the relevant parameters, is just the product of the probabilities of each outcome separately. Jarrett contends

that the evidence implies only that at least one of these two conditions is violated, but only the former condition is entailed by special relativity. On the basis of this, it has been argued that parameter independence is to be retained, and outcome independence must be given up. In particular, it is usually believed that a rejection of the latter would not determine the possibility of superluminal signalling, which is what relativity rules out, because whenever outcome independence fails, an element of randomness is present which makes superluminal signalling impossible. As Jarrett puts it, if quantum systems only violate outcome independence, no contradiction with relativity arises, as

> it is a consequence of the failure of determinism that measurement outcomes are not (even in principle) under the control of experimenters. [1989; 77]

However, it is unclear whether the fact that correlations cannot, in practice, be exploited for superluminal signalling is sufficient for claiming that relativity is safe. It might, instead, be maintained that relativity forbids *any* type of non-local connection, not only the transmission of humanly exploitable information between space-like separated regions. Moreover, it has also been argued that, even if quantum systems only violate outcome independence,

> the possibility remains open that the experimenter might use some controllable feature of the experimental situation as a 'trigger' which operates stochastically on the outcome at her own end of the experiment. The signaller could then influence, without completely controlling, the result in the individual case, and could thus signal superluminally by employing an array of identically prepared experiments. (Jones and Clifton [1993; 301])

In addition to this, that relativity prohibits superluminal signalling has itself been put into doubt (Friedman [1983; Secs. 4.6–4.7]). In between the extremes, a large number of positions have been taken. For instance, Fine [1989] denies that the detected correlations need an explanation at all; while Winsberg and Fine [2003] suggest that the joint state could, in fact, be wholly determined by the separate states of the two particles, although by a functional relation other than multiplication, and the correlations consequently be perfectly explicable in local terms. The entire 'Jarrettian' approach considered so far has also been questioned: Maudlin, for instance, argues that if the aim is to study the nature of quantum non-locality, it is misleading to perform general analyses of the statistical (in)dependence between the parts and the whole, and it is instead necessary to look at the specific ontologies postulated by the various interpretations of quantum mechanics directly [1994; 94–98]. It should be evident, then, that – while the evidence is obviously taken for granted – what revisions experimental metaphysics leads us to make in our beliefs and hypotheses is far from clear, and definitely does not obey just logical rules.

2. A second important point has to do with the nature of H, that is, of the metaphysical hypothesis that one is supposed to put to the test. Sceptics about metaphysics could claim that locality is not really a metaphysical hypothesis, and it is instead a very general empirical claim that follows from our repeated experience of the world and from its being sanctioned at a higher degree by a well-confirmed and successful scientific theory. To the extent that this is not just a terminological issue, those who, like Shimony, feel entitled to talk about experimental *metaphysics* should then provide a positive definition of what counts as a metaphysical conjecture in the first place. While nothing in this direction can be found in the literature on quantum correlations, however, our discussion in the previous chapter should suffice to fill this gap. In light of that discussion, in particular, that locality truly is a metaphysical hypothesis might indeed be questioned: at a first glance, at least, Einstein's statement of locality by no means uses concepts and terms external to physical science; and, indeed, the principle appears to be a direct consequence of relativity theory and nothing else. To claim that, nevertheless, the principle became part of relativity theory insofar as it was part of the metaphysical presuppositions that were in play at the moment in which the theory was devised only pushes the issue one level back. The impression still lingers that, at some point, what counts as metaphysics must be defined non-circularly and without regress. Here, it will be assumed that (some degree of vagueness being perhaps inevitable) our proposed criterion based on the non-eliminable use of philosophical concepts and categories is appropriate for this purpose. On that basis, it can be plausibly be argued that it is not a) locality per se, but b) its individuation, by *a priori* means and using the specific tools of philosophical analysis, as one of the possible ways the world might be, that counts as metaphysics.

Be this as it may, on the other hand, the important fact remains that what is relevant in the discussion of locality in the quantum domain is the way in which empirical data and scientific theories are brought to bear on one's attempt to evaluate and assess theoretical hypotheses. Regardless of the nature of these hypotheses in the case that is taken to be paradigmatic of (if not identical to) experimental metaphysics, i.e., the case of quantum correlations, the same kind of reconstruction suggested above seems available for truly metaphysical hypotheses whenever they are at least potentially connected to specific scientific domains (which, it was argued earlier, typically happens when the former can be used to ground the interpretation of the theories describing the latter). The dynamics pointed out by Shimony, that is to say, does, indeed, seem to capture an important aspect of the proper naturalisation of metaphysics, related to what we called the indirect testability of metaphysical hypotheses, i.e., their use for the interpretation of empirical data as they are described and made sense of by our best science.

3. The third and last point that must be made concerns another element that is left completely undefined in the extant literature on experimental metaphysics. If, as illustrated a moment ago (point 1), experimental metaphysicians do not just apply *modus tollens* but, rather, take the empirical evidence to require some sort of complex reassessment of *both* the (allegedly) metaphysical hypothesis under scrutiny and of the theory used to devise a test of that hypothesis, then how this reassessment should be carried out must be explicitly stated. As the brief illustration of the options entertained in the literature on quantum correlations offered earlier illustrates, doing this is by no means easy, for – as things stand – what criteria one should apply is definitely unclear. This means, obviously enough, that it is of paramount importance for the naturalistic metaphysician to try to define these criteria, and thus the very methodology of constructive naturalism about metaphysics, in as much detail as possible. An attempt in this direction will therefore be made next.

3.   Theory-choice in (naturalised) metaphysics

What principles (if any) can and should guide us in evaluating and selecting metaphysical hypotheses in the light of the indications coming from science, so also providing the most plausible interpretative background for scientific theories themselves?

As is well known, criteria for theory-choice in the case of science have long been identified and widely discussed in the past, and it is agreed that they are essentially the following, well-known at least since the work of Kuhn: empirical accuracy, logical consistency, breadth of applicability, simplicity and fruitfulness. True, the debate about their status and the possibility to employ them in an objective way is open. Still, if there are pragmatic criteria of theory-choice in science, it is agreed that the list includes these. In the case of metaphysical theory-choice, *prima facie* it seems that these criteria can be preserved, albeit with some modifications. For instance, empirical accuracy cannot, of course, be interpreted literally as direct fit with experimental data and observations. Rather, it must be intended, more loosely, as compatibility with those data. In this sense, for instance, presentism (the theory according to which – very roughly! – everything that exists occupies a unique instant of time) appears – at least at a first glance – not to fare too well on this score. For, relativity theory is normally taken to pose a clear challenge to such a view. We will discuss this issue in more detail later (Chapter 4), but for the moment the idea of *prima facie* (in)compatibility between a metaphysical hypothesis and the relevant scientific theory (or theories) should appear sufficiently intuitive for our purposes. Analogously, in the case of metaphysics, fruitfulness cannot be intended literally in terms of new predictions or unification of independent phenomena, but should instead be understood in terms of relevant new consequences at the theoretical level, first and foremost in the form of interpretative claims. For example, the metaphysical interpretation of a theory x might turn out to have a (n *a priori* identifiable) consequence that naturally fits with the indications coming from another theory y which was not taken into account in the first instance. This sort of unexpected agreement might appear at the purely metaphysical level, for instance, when a metaphysical hypothesis turns out upon analysis to fit with or provide a basis for (a specific consequence of) another metaphysical hypothesis previously regarded as independent of the first. But all sorts of trans-theory connection in science and metaphysics appear possible.

However, there is more to say. Going back for a moment to the presentism/relativity example, it might be thought that one of the things that are obtained by switching to a naturalistic metaphysics is the possibility to *conclusively* discard certain metaphysical options (and regard others as certainly correct) based on the empirical data – in this case, presentism based on relativity theory. But this is not so (even though, of course, one cannot be entirely sure that this *never* happens – this is not the point here). For, there is (almost) never a direct relation of logical entailment between a scientific theory T and a metaphysical hypothesis M. (Indeed, this is an important reason why metaphysics must be preserved as an autonomous enterprise alongside science: metaphysics simply cannot just be read off from our best science!) Rather, once all the relevant factors are taken into account, a theory M' which *prima facie* looked less plausible than M might turn out to be the best available basis for the interpretation of T. In other words, the criterion of compatibility with the empirical data cannot be intended as something with a mechanical application. Instead, it must be applied in parallel with the other criteria, and its indications put into a systematic relation with those coming from those other criteria. Crucially, such criteria do not only include those suggested by Kuhn. Something else must be added, which is in general not given due attention (essentially, because of a basic misunderstanding of its content and significance). The best work to refer to here is that of Quine and Ullian (especially [1978; chapter 6]). Emphasising how – as every philosopher knows – our universal generalisations based on empirical data cannot be conclusively validated on the basis of deduction from self-evident truths and observation, Quine and Ullian argue that the beliefs that we entertain about those generalisations can in any case be rationally justified, and indeed need to be. A non-deductively formed belief is rationally justified, they argue, whenever it is formulated while taking into account (implicitly or explicitly) a number of essential virtues that no successful hypothesis about the world can lack. Quine and Ullian propose five such virtues: generality (corresponding to Kuhn's breadth of applicability); simplicity (to be intended as minimisation of complexity) and modesty (corresponding to logical weakness and dependence on plausible assumptions) – these two together, more or less, coincide with Kuhn's simplicity; refutability (a requirement of clearly Popperian flavour, to be conceived of, in the case of metaphysics, not in terms of direct falsifiability but, rather, in terms of (additional) critical assessment based on new metaphysical conjectures and empirical data). In addition to this, and this is the key point here, Quine and Ullian emphasise the role of the virtue of *conservativeness*. To make this clear from the outset, their idea is NOT that one should be conceptually conservative at all costs, let alone always trust less revisionary hypotheses. (This is the kind of misunderstanding that seems to have led, and still lead, many philosophers to simply ignore this point and instead accept all forms of conceptual revision, independently of the actual need for them.) Rather, the idea is the following. First of all, whenever we want to explain something new, we introduce a new hypothesis. Such hypothesis might conflict with beliefs that we already have. Since acceptance of the new hypothesis (necessary for the sought explanation) implies acceptance of its consequences, our need for explanation entails the possibility of conflict between old and new beliefs. Therefore, some revision in our web of beliefs is required, and our aim must be (obviously enough) to eventually obtain a new web of beliefs which is consistent and includes the new explanatory hypothesis. In this context, Quine and Ullian contend, changing as little as possible while obtaining this latter result is not only advisable *but also necessary*, given the amount of conceptual work needed as well as the fact that new explanations are continuously sought and, consequently, new adjustments and conceptual revisions always required. Compare: if we are used, say, to getting on the bus at the same time every morning and being successful in doing that, and on a certain day our expectations in this sense turn out to be incorrect as the bus does not show up when expected, our first reaction is not to consider as the most plausible the hypothesis that we were completely wrong about the bus' timetable until the surprise day. Rather, we stick – conservatively – to the hypothesis that there is an objective timetable for the bus, which we have been successful in getting to know and use so far; and expect there to be a more 'local' explanation of the specific 'recalcitrant' fact about the bus, one that does not fit the general scheme and yet can be accommodated within it relatively easily: the bus might have had an engine failure. In Quine and Ullian's words then, we should (to repeat, without in any way taking this to point to a more extreme conservatism, according to which revisionary new hypotheses about the world should in principle be looked at with suspicion) obey the following rule:

In order to explain the happenings that we are inventing it to explain, the hypothesis may have to conflict with some of our previous beliefs; but the fewer the better. ... The less rejection of prior beliefs required, the more plausible the hypothesis – *other things being equal*. [Ib., italics added]

According to Quine and Ullian, indeed,

Conservatism is rather effortless on the whole, having inertia in its favor [and it...] holds out the advantages of *limited liability* and a *maximum of live options for each next move*. [Ib.; italics added][25]

The final part of the last quotation is particularly important. That the way of thinking recommended by Quine and Ullian is *not* prone to favour obstinate entrenched beliefs in the face of contrary evidence is suggested by the fact that it was openly endorsed by a pragmatist like William James. In his *Pragmatism* [1907], James first describes and advertises the pragmatist method as representing the empiricist attitude in its best form; that is, one not concerned with first principles and supposed necessities, but rather with consequences of specific assumptions, particular things and individual facts. After having taken this stance – which,

it would seem, modern-day empiricists *à la* Van Fraassen or other detractors of Scholastic metaphysics such as Ladyman and Ross should find
congenial – James also explicitly states that often new facts "oblige a
rearrangement" in the set of beliefs that we take to be true, and thus to
correspond to objective facts. And, at that point, he claims that

> by modifying his previous mass of opinions [... one faced with new
> evidence that he wants to integrate with what he already knows]
> saves as much as he can, for in this matter of belief *we are all extreme
> conservatives*. So he tries to change first this opinion, and then that
> (for they resist change very variously), until at last some new idea
> comes up which he can graft upon the ancient stock *with a minimum
> of disturbance of the latter*, some idea that mediates between the stock
> and the new experience and *runs them into one another most felicitously
> and expediently*. [1907(1979); lecture II 'What Pragmatism Means',
> italics added]

Indeed, according to James,

> The most violent revolutions in an individual's beliefs leave most of
> his old order standing. ... We hold a theory true just in proportion to
> its success in solving this 'problem of maxima and minima'. ... Their
> influence [that of the older truths] is absolutely controlling. Loyalty
> to them is the first principle – in most cases it is the only principle.
> [And ...] new opinion counts as 'true' just in proportion as it gratifies
> the individual's desire to assimilate the novel in his experience to his
> beliefs in stock. [Ib.]

Of course, the pragmatist will not take the most established beliefs as
untouchable. Rather, it will be ready to entertain any hypothesis as long
as it is the most useful and expedient one, once the entire web of beliefs
and the various ways in which it could be modified have been considered. It should thus be evident by now why it was claimed earlier that the
virtue of conservatism being defended here is entirely compatible with
the most radical and revisionary hypothesis (first and foremost, in our
scientific account of empirical evidence) being accepted as the most likely
to be true. Indeed, the sort of conservatism being advocated here does
not concern *what* new beliefs can/have to be added to our existing 'stock'
given the evidence, but rather *the way in which* this should be done.[26]
    Another thing that should not at this point look like an implausible
form of rigidity is the view that, on the construal being envisaged,

common sense becomes as an important reference point in the process of revision. In fact, it seems natural to think that, other things being equal, it will be only normal for us to insert a new belief (as revolutionary as it might be) in our existing web of previous beliefs by changing as little as possible, so preserving as much of our common-sense beliefs as possible. As is well-known, common sense was strongly defended as a starting point of philosophical inquiry by Reid (see for instance, his [1785 (2002)]) – who, at any rate, took it to constrain rather than command philosophical theses. Some time later, another American pragmatist, C.S. Peirce, similarly adhered to what he called 'Common-sensism' – so assuming the existence of indubitable propositions and inferences – but also made it 'critical' – so regarding the indubitable in any case subject to further reflection (for a short discussion, see Peirce [1905]).

All this may sound well-known, if not obvious. And, indeed, the emphasis put on conservativeness and common sense is certainly not new. However, it is believed here that, first of all, a clear understanding of why and how one should be conservative is of fundamental importance. And, secondly, that the specific way in which the various authors mentioned above look at conservativeness and common sense and put them in relation to the evidence on the one hand, and to other theoretical virtues and pragmatic criteria on the other is very indicative and functional to our present purposes. In a way, the proposal that is being put forward attempts to improve on Sellars' [1962] well-known idea of seeking a 'stereoscopic' mixture of the scientific image and the manifest image of the world. The idea is, in particular, that metaphysics and science together can, first, define a more complete and precise scientific image; and that, secondly, a liberal form of naturalism developed along the lines suggested so far is the best tool to determine how and to what extent such an image truly departs from the manifest image, and how the sought stereoscopic vision is to be achieved.

Let us take stock before continuing our discussion. The situation is as follows: since there is no interpretation ready to be extracted from science, nor is the 'fusion' of science and metaphysics in any way an automatic process, the naturalisation of metaphysics should be understood as entailing the application of a number of pragmatic criteria for theory-choice; among these, in particular, a principle of conservatism or 'minimal revision' – giving at least initial privilege to common sense and established beliefs over radical new hypotheses – seems to play an important role that cannot be neglected in reading our scientific theories philosophically. It is because of this dynamics that hypotheses that might *prima facie* look like bad candidates for complementing and making sense of scientific data (think, again, of the presentist metaphysics in relation to Einsteinian relativity) might eventually turn out to be preferable with respect to the existing alternatives (in this case, for instance, a conception according to which the past/present/future distinction is relative to frames of references or a conception according to which becoming and/or time do not exist at all). For, it becomes possible that such hypothesis is, after all, the best one in virtue of its greater harmony with common-sense intuitions – *other things being equal*; or that it turns out to fare better than the alternatives when all the criteria and theoretical virtues have been considered together. Clearly, a careful evaluation of all the relevant factors is crucial here, and it is not obvious that procedures for quantifying the parameters to be taken into account and then comparing the various alternatives in anything like an objective way are available. But why should criteria of theory-choice only be applicable to the extent that their respective weights can be precisely quantified? After all, what we are looking for are *some* criteria for the evaluation of hypotheses and the critical evaluation of alternative options. That these criteria should lead to uncontroversial, objective and shared conclusions seems to be an additional – albeit important – request, the lack of which does not entail the collapse of the entire project. Indeed, a similar lack of an objective 'measure' certainly does not entail – at least not in an obvious way – that talk of pragmatic criteria and theoretical virtues should be given up in the case of scientific theory-choice. Why should the same not apply when it comes to putting science and metaphysics together?[27]

An objection (voiced, for instance, by Ladyman [2012]) might be that, *unlike* in scientific theory-choice, one cannot, in fact, have recourse to theoretical virtues and pragmatic criteria for assessing metaphysical conjectures, not even when these are systematically looked at (also) from the point of view of scientific theory. The reason for this would be that the cost-benefit analysis in terms of these criteria and virtues does not lead anywhere in metaphysics, because metaphysical theories, in addition to having global rather than local application, are *strongly* and not just *weakly* underdetermined by empirical evidence. That is, they are *underdetermined with respect to all possible observations* and not just all observations carried out until now. In view of this, Ladyman argues that we have inductive grounds for believing that pursuing theoretical virtues such as simplicity and the likes is advisable in science, but no such thing holds for metaphysics. Indeed, says Ladyman, the historical development of science seems, in itself, sufficient to show that underdetermination is not a real problem for science (i.e., a problem for the practice of science and for a realist reading of it) but things stand differently in the case of metaphysics, where theoretical alternatives essentially proliferate without any form of effective selection.[28]

The response to this is that, here at least, the suggestion being made is not that theoretical virtues are exactly as important and as truth-conducive in metaphysics as they (perhaps) are in science. We also put a fundamental emphasis on the idea that those theoretical levels which are more abstract and farther away from evidence and empirical data must be supported by being systematically put into relation with other levels, typically scientific ones, which are closer to such data. It is in this sense that, in the previous chapter, it was argued that although metaphysical conjectures are not directly testable, they are, and must be, at least indirectly empirically relevant. What this means is exactly that the seemingly strong underdetermination besetting metaphysics might (albeit perhaps not in all cases) be shown to be weak after all, as conjectures that previously seemed entirely abstract and disconnected from reality (may) turn out to make a difference when it comes to interpreting specific theories that certainly have an empirical basis. After all, what principled way could there be for determining when a given hypothesis is really strongly underdetermined? Would this not be like claiming that, while mathematics is generally useful for the development of physics and so the pursuit of mathematics for its own sake is generally justified, there is a specific bit of mathematics that is, in principle, useless for physicists? Or, worse, like saying that mathematics should not be pursued because abstract, unless somehow already proven to be useful for empirical scientists at every step of its development? This is by no means to be intended in the sense that metaphysical and scientific hypotheses are on a par with respect to their responsiveness to empirical data and criteria for hypothesis selection more generally. This is exactly the starting point of the form of naturalism being proposed here.

In conclusion, there is hope that putting an emphasis on empirical adequacy, on the one hand, and conservativeness (primarily, with respect to common sense) on the other, can at least alleviate the worries that have been manifested with respect to pragmatic criteria of theory choice in the past, both in general and in the case of metaphysics. It is in this sense that the proposal formulated in this section is believed to be more than just a reiteration of well-known approaches to theory-choice, whose limitations have already been pointed out clearly.[29]

4. Conclusions

In this chapter, we have attempted to define more precisely the sort of constructive naturalistic approach that, it is claimed, should be implemented when trying to do metaphysics in a way that pays due attention to the unique significance of the empirical sciences. It has been argued that such an approach should be based on the view that metaphysics is the study of a *sui generis* possibility space (which does *not* require the autonomy and irreducibility of metaphysical modality), in particular with a view to identifying essential features of (metaphysical) kinds of things, involving their identities and their mutual relations of priority and dependence.[30] In this perspective, as suggested by supporters of experimental metaphysics but in a way that is much more complex than it may appear from the related literature, from an initial dynamics of *modus tollens* involving metaphysical hypotheses, scientific theories and the empirical consequences of the two considered together, naturalistic metaphysicians should derive the material for their peculiar sort of inquiry.[31] The latter involves evaluating the various kinds of adjustments that might suffice for making the pertinent metaphysical hypotheses fit with the relevant scientific data, and the various ways in which the emerging alternatives fare with respect to criteria for theory-choice. Arguments have been given to the effect that, when it comes to the latter, a moderate form of conservatism is advisable, which does not entail a principled refusal to update our worldview (even in radical ways) on the basis of incoming evidence; and to the effect that, in the form it was presented, naturalism can effectively circumvent the problem of radical, strong underdetermination of metaphysics by the evidence (that is, essentially, the problem that the likes of Putnam and Van Fraassen – as illustrated in the previous chapter – consider fatal for contemporary metaphysics).

The resulting picture seems to be in many ways akin to the sort of 'liberal naturalism' advertised in recent literature, for instance, in the collection of essays edited by De Caro and Macarthur [2010]. Liberal naturalism promises to overcome the problems that seem to arise from a more narrow naturalism that makes the *ontological* claim that the world consists of nothing but the entities to which successful scientific explanations commit us and/or the *methodological* claim that all forms of knowledge are either illegitimate or reducible in principle (or, of course, identical) to scientific knowledge. This is done by claiming that naturalists need not be ontological/methodological reductionists, and that continuity between science and philosophy can and should be intended more loosely.[32] This is exactly what we have tried to show so far in this book.

With all this in mind, it is now possible to move on to the application of the proposed naturalistic methodology to some case studies. As pointed out in the introduction, these concern physics but – although it is believed that physics does, indeed, occupy a privileged position among the sciences in terms of priority and fundamentality – this is due to personal interest and preferences rather than physicalist/reductionist presuppositions. We will begin, in the next chapter, with a discussion of the ontological commitments (first and foremost, with respect to the notions of object and individuality) suggested by quantum physics (primarily, non-relativistic quantum mechanics).

Notes

1. In particular, they point out, "Philosophers have often regarded as impossible states of affairs that science has [then] come to entertain" [Ib.; 16], and provide the examples of non-Euclidean geometry, indeterministic causation and non-absolute time.
2. Philosophical analysis might lead one to realise that there is (perhaps temporarily) only one conceivable possibility in a certain case; or that something is impossible due to internal inconsistency. Then, clearly, it would go beyond the 'might (not) be' and say something about the 'is (not)'. Also, metaphysicians typically argue in favour of one specific hypothesis rather than just indicating a few of them. This, however, does not affect the main point, which is that metaphysicians typically deal with the whole range of alternative hypotheses about certain aspects of reality, and logic is normally insufficient for selecting just one among these alternatives.
3. It is also worth reminding the reader of something that has been nicely remarked by Monton [2011]. According to Monton – as mentioned in the previous chapter – not only is the literal falsity of scientific theories and their occasional mutual inconsistency a problem for the scientific realist and, consequently, for the naturalised metaphysician; but it is also the case that, if one looks at historical examples of conflicts between science and metaphysics, it is not obvious that one should bet on the (then-) current science (one of Monton's examples concerns the Cartesian objection to Newtonian gravity based on a requirement of continuity which used to be of merely philosophical relevance and is nowadays regarded as also scientifically respectable).
4. We will use these terms interchangeably from now on.
5. Some definitions and additional considerations are probably useful at this point. *Metaphysical* possibility/necessity is truth in some/all metaphysically possible world; *physical* possibility/necessity is logical consistency with/

being entailed by physical laws, i.e., those statements, appearing in current scientific theories, that describe necessities in the actual world (scientific realism being assumed, of course). In addition to these, there is *doxastic/epistemic* possibility/necessity: roughly, whatever x a subject S, given what S knows, does not rule out/rules out as impossible when x is negated; *logical* possibility/necessity, that is, consistency with or derivability from axioms within a logical system (typically, classical logic); and *conceptual* possibility/necessity, that is, compatibility with, or being the negation of, something incompatible with the set of all conceptual truths (i.e., statements such as 'All triangles have three sides'). Clearly, metaphysical modality cannot just coincide with logical modality in its strict sense: the constraints set by logic alone are too weak for delimiting the kind of information we expect our metaphysics to give us. For instance, it might be true in all (metaphysically) possible worlds that properties are universals, but certainly nominalism about properties is not logically impossible. Additionally, notice that there are various systems of logic but, it would seem, only one family of metaphysical possibilities/necessities; and also that there are '*im*possible' worlds, constructed on the basis of paraconsistent logics. (It can instead be contended that broad logical possibility/necessity *is* metaphysical possibility/necessity, see for instance Hale [1996]). Similarly, metaphysical modality cannot reduce to epistemic modality, for it seems that what is metaphysically possible/necessary must outstrip what is deemed possible/necessary by specific subjects. For all these reasons, the focus must primarily be on the relationships between metaphysical possibility/necessity and conceptual and physical possibility/necessity.

6. The following is perhaps worth pointing out. The view has been put forward – among others, by Shoemaker [1998], Smith [1996], Swoyer [1982], Ellis [2001] and Mackie [2006] – that laws of nature fully qualify as metaphysical necessities. Even taking these as arguments for the co-extensiveness of natural and metaphysical necessity, which is already a stretch, they are different from the argument being scrutinised here: rather than suggesting that we do not need to posit a kind of modality over and above physical modality, these authors 'push' the latter, allegedly 'lower', level up towards the former.

7. It must be pointed out that there are alternatives, based on either understanding (see, for instance, Bealer [1999] and Peacocke [1998]) or counterfactual reasoning (Hill [2006], Williamson [2007]). However, it is far from obvious that these provide clear advantages over conceivability-based accounts, or even that they do not, in fact, presuppose modal notions. For a claim to this effect with respect to Williamson's counterfactual-based accounts, for instance, see Tahko [2012]. At any rate, even independently of this, focusing on conceivability will suffice for our present purposes.

8. Incidentally, Yablo also explicitly allows for defeasible justifications of conceivability/inconceivability claims, which points to the fallibility of metaphysical statements – something that we have already deemed desirable, if not necessary, in a context of naturalistic metaphysics. Yablo's approach was further developed by Geirsson [2005]. A relevant, related view is that according to which metaphysical possibility coincides with ideal positive conceivability (see Chalmers [2002]), but we do not need to add more details to the discussion in the main text.

9. For instance, consider (again), Leeds' view that 'Water is H_2O' is, in fact, a physical necessity such that we cannot conceive that it is physically possible for 'water' not to refer to (the kind essentially defined on the basis of its being composed of molecules of) H_2O. To be sure, to regard that expressed by 'Water is H_2O' as an *a posteriori* necessity we just need empirical facts (that examined token instances of the kind water were – scientific realism! – composed of H_2O molecules) and conceptual truths (i.e., that natural kinds share essential structural properties across their token instances). But we also need (i) the concept of a natural kind and, most importantly, the assumption that (ii) natural kinds exist (and iii) to have interacted with one in the case of water).

10. With respect to physical necessity, Fine basically argues that some natural necessities are metaphysically necessary (e.g., electrons have negative charge: nothing would be an electron while failing to have negative charge) and some are not (e.g., light has maximum velocity c, for the value of c in the actual world: light would still be light in a possible world where c is different). Fine also argues that natural necessity cannot be regarded as a relativised form of metaphysical necessity, because there seems to exist no such relativisation that does not make the relevant necessities trivial or, at any rate, way less substantial than we perceive them to be.

11. Essentially, the point is that, even taking metaphysical necessities to be those natural necessities that are essential truths, one fails to capture the relevant modal force, with respect to which metaphysical modality intuitively appears to be more fundamental.

12. In response to those monists (e.g., Jackson [1998]) who consider conceptual modality fundamental, instead, the arguments in favour of the irreducibility of the metaphysical vocabulary provided above can be employed again. Incidentally, in the context of this sort of monism, whether metaphysical questions are merely internal questions in Carnap's sense becomes a particularly interesting question. Whether or not the Carnapian perspective is endorsed, there is wiggle room for the defender of metaphysics: even if s/he grants that metaphysics is essentially a framework, he or she can claim that such framework is the most encompassing one available and 'absorbs', as it were, all others.

13. The latter option appears more convincing. But a choice in this sense is not required in what follows: as argued in the main text, constructive naturalism about metaphysics is compatible with both views.

14. Callender also complains that metaphysical *equivalence* has not been defined so far, and indeed appears to be an obscure notion. However, it is not clear why this should be so. In fact, some work towards providing a thorough definition of metaphysical equivalence has already been done. Miller [2005], for instance, argues that metaphysical equivalence amounts to correct inter-translatability and provides criteria for diagnosing whether translations are correct (existence of an assertibility mapping that is truth-preserving, plus equality of explanatory power, empirical equivalence and sameness of theoretical level).

15. Fine's notorious example is that if essential properties are those possessed necessarily, then it is an essential property of Socrates that he belongs to the singleton whose sole member is Socrates, but this seems incorrect. Fine's proposal is thus to regard essences as fundamental, and as having to do with the things' identities and existence. (As a consequence of this, metaphysics

becomes primarily a search for 'real definitions' of things rather than for basic modal truths). There are replies available to Fine's claims, but his approach appears entirely justified from the present perspective, aiming to provide a characterisation of metaphysical possibility and necessity claims themselves.

16. It is not surprising, then, that Lowe himself raises doubts about the actual strength and significance of arguments along the well-known Putnam/ Kripke line. Hopefully, the remarks just made in the main text will also justify the choice of not entering the discussion concerning whether essences are accessible a priori (Lowe [forthcoming]) or a posteriori (Oderberg [2011]). If pushed, the constructive naturalist should probably opt for the latter option.

17. Schaffer traces this sort of approach back to Aristotle's study in the *Metaphysics* of substances as those entities that are fundamental and such that everything else depends on them. For the time being, we are still using notions such as dependence or reduction in a somewhat vague and loose manner. Things will become more precise shortly.

18. Without assuming that in all cases the claim that fact(-type) x is grounded in fact(-type) y – that is, x obtains in virtue of y – entails the ontological reduction of x to y.

19. See Schaffer [2009; 366–373]. As for the last point, in particular, Schaffer notices that the Quinean criterion leads to commitment to the existence of fundamental entities plus grounding relations and grounded entities. This is the only way, for instance, in which fundamental physics can be taken seriously in what it tells us about particles or fields while resisting eliminativism with respect to things, such as table and chairs, the existence of which is dependent on that of those particles and fields.

20. In Schaffer's words: "Existence questions do play a role for my sort of neo-Aristotelian. What exists are the grounds, grounding relations, and the grounded entities. Hence, existence claims constrain the grounds and the groundings, to be basis enough for the grounded" [2009; 353].

21. This is mainly intended to exclude relations involving abstract entities, such as the relation between a physical fact and the proposition it makes true – which might, indeed, be one of ontological dependence. A thorough discussion of ontological dependence can be found in Lowe [2010], who interestingly regards it as a form of essential dependence, primarily having to do with the identity of things, i.e., with what makes them the things they are. See also Correia [2008].

22. Here, one might disagree on the basis of something like Lewis's doctrine of Humean Supervenience (see, e.g., Lewis [1986b; ix]) and the view that it captures the fundamental structure of the world. But the point here is deeper: supervenience relations are mere relations of modal correlation and have the wrong formal features for capturing the intuitions underpinning the idea of fundamentality: they are reflexive and non-asymmetric while dependence claims appear irreflexive and asymmetric; and they are intensional rather than hyperintensional, so becoming vacuous for necessary entities. On this basis, one can contend that what Lewis had in mind really was more than supervenience, i.e., roughly, the dependence of wholes and their properties on parts and their properties.

23. For specific defences of the notion of grounding, see Raven [2012] and Audi [2012]. For discussions, Rosen [2010] and Trogdon [forthcoming]. Among the issues that have not been discussed here, one can mention: whether ground is binary or polyadic; whether it only connects actual facts or also merely possible facts; the distinctions between full and partial, mediate and immediate and weak and strict ground; the nature of the relevant semantics. Fine [2012] usefully discusses these topics as well as the connection between grounding and essences (discussed in the previous sub-section).

24. It is very important to stress that the violation of Bell's inequalities has been confirmed experimentally and thus does not just follow from quantum mechanics at the abstract theoretical level. Experiments reproducing scenarios sufficiently analogous to that described in the main text, in particular, have been carried out as early as 1972 and 1976. The most famous of these were those performed by Alain Aspect and collaborators (see, e.g., Aspect, Grangier and Roger [1981] and [1982]).

25. Of course, all this is essentially Quine's [1951] idea of 'minimum mutilation' of existing beliefs, only connected to other criteria for updating beliefs as well, and also extended beyond Quine's original discussion of the web of beliefs and general belief-types (for instance, logical, mathematical, physical).

26. It is interesting to notice that exactly the same dynamics is taken for granted in the completely different domain of the logic of belief revision. There, a set of beliefs K can be *expanded* so as to contain a new belief p, *contracted* so that a belief is removed from it, and *revised*, in which case, a sentence p is added to K, and at the same time, other sentences are removed as needed to ensure that the resulting belief set is consistent. Now, revision is essentially an expansion that follows a prior contraction by all beliefs inconsistent with the new one(s). And it is a basic assumption of the framework that belief contraction should not only be successful, but it should also be minimal in the sense of leading to the loss of as few previous beliefs as possible. Epistemic agents should, in other words, give up beliefs only when forced to do so, and should then give up as few of them as possible. (For more details, see Hansson [2011]). For an articulation of the idea that the most entrenched beliefs – i.e., those that are most useful in inquiry or deliberation – should 'weigh more', see Gärdenfors [1988] and Gärdenfors and Makinson [1988]. For critical discussion, see Rott [2001] (who connects extant work on belief revision with Quine's views on belief change in a way that is clearly relevant for our purposes).

27. A very similar view is defended by Paul [2012]. Paul argues that metaphysics is essentially an activity of model-building (on this, see also Godfrey-Smith [2006]) based on general categories of things, and that "[m]etaphysical theories, like other theories, compete with respect to their theoretical qualities as well as their empirical adequacy [...and] if we accept inference to the best explanation in ordinary reasoning and in scientific theorizing, we should accept it in metaphysical theorizing" [Ib.]. Paul also agrees that "tradeoffs between maximizing theoretical virtues (e.g., simplicity and elegance) and preserving commonsense beliefs are common, but tradeoffs that sacrifice central commonsense beliefs only rarely convince" [Ib.].

28. In discussing Paul's work mentioned a moment ago, Ney [2012] expresses the similar worry that metaphysical hypotheses have very little initial

confirmation, and so theoretical virtues cannot do much to help with theory-choice, and more substantial empirical backing is required.

29. Think, for instance, of Lewis' cost-benefit analysis in favour of modal realism [1986; 133–135], often looked at with scepticism both because of its conclusion and of the relevance given to elements such as simplicity and explanatory power. Just to mention one thing, in Lewis' analysis, empirical adequacy and conservativeness did not (and could not, it would seem, given the nature of the issue dealt with by Lewis as well as his general approach) play any role.

30. The resulting view is analogous to that proposed by Fine [2012a], who claims that there are "five main features that serve to distinguish … metaphysics from other forms of enquiry. These are: the aprioricity of its methods; the generality of its subject-matter; the transparency or 'non-opacity' of its concepts; its eidicity or concern with the nature of things; and its role as a foundation for what there is" [Ib.; 8]. We have not discussed the transparency feature (i.e., roughly, the idea that metaphysical concepts directly and unambiguously relate to and signify the things they are concepts of, by the very nature of the field they belong to), but the general perspective is clearly analogous.

31. Or, to be more precise, for the second part of their work, which follows a fundamental prior exploration of possibility space, that is, the preliminary formulation of purely metaphysical hypotheses.

32. In an interesting paper in the collection, De Caro and Voltolini [2010] respond to the following dilemma for liberal naturalists formulated by Neta [2007]: either liberal naturalism coincides with ontological 'canonical' naturalism, or it acknowledges the existence of items which are completely irreducible to those posited by science. In the former case, though, it is not a fully autonomous position, while in the latter it is, in fact, not a naturalistic position after all. In response to this, De Caro and Voltolini convincingly point out that it is possible to posit entities (such as moral, abstract or intentional entities) which, while correctly accounted for only through forms of understanding clearly different from, say, physics, neither require one to include them in the range of what is relevant for science (say, the causal relations explored by physics) nor are to be counted as supernatural. The characterisation offered here of categories and concepts such as universals or presentism as typically metaphysical goes exactly along these lines: these categories and concepts are significant to the extent that they concern domains of reality we can gain access to, and knowledge of, by empirical means. Yet, they need not (and most likely cannot) be reduced to categories and concepts of the sciences. (As argued earlier, this appears compatible with both pluralism about modality and a monism that gives priority to nomological modality.) In connection to this, it is worth mentioning that De Caro and Macarthur themselves, in a recent edited collection of papers by Putnam [2012], provide arguments to the effect that Putnam himself has lately become a supporter of liberal naturalism, while also revising his claims about the uselessness of certain philosophical questions, which we discussed in the previous chapter (see the introduction of the collection, and also essays 5 'The Content and Appeal of Naturalism' and 28 'Wittgenstein: A Reappraisal'). To the extent that Putnam seems

to prefer liberal over narrow scientific naturalism because of the seeming irreducibility of normative facts, moreover, it is also worth noticing the link between him and Fine's [2002] modal pluralism (metaphysical, physical and normative modality as distinct and irreducible), mentioned earlier in this chapter.

3
Matter

1. Identity and individuality in quantum mechanics

Although it has been intended in different ways in the literature, individuality will be assumed here to simply consist in the possession of determinate self-identity and numerical distinctness from other things (synchronically and, possibly, diachronically – only the former will be our focus here). To distinguish it from particularity, it will be assumed to be a feature of property-bearing objects provided with a certain internal unity and structure: the redness of an apple (be it a trope or an instance of a particular) will thus be a particular, while a red apple (provided that it has not been, say, smashed into scattered pieces) will be an individual.[1] It is, of course, an issue whether something should be added to this (for instance, being a substance in the sense of being independent of other things with respect to existence and identity), but the above definition is the most general and the most neutral with respect to a range of open metaphysical questions – in particular, with respect to the debate about identity and individuality in quantum physics to be explored shortly.

It has often been argued that individuality can be reduced to something more fundamental. The idea that an entity is an individual if and only if it is *indivisible*, for instance, was defended by Saint Thomas Aquinas in his *Summa Theologica* ([Part I, Question 11, Article 2]). This view was later refined by Suarez, who argued that individuality is indivisibility into entities *of the same specific kind* as the original one. Further specifications can be added, as illustrated, for example, by Gracia, who considers the possibility that individuality is indivisibility into entities of the same *quantity* as the initial one [1988; 29–32]. The view is, in any case inadequate, however, as whether or not an entity possesses the identity conditions defined above for individuals is logically independent of whether or not it has parts (of a certain type). At best, this conception manages to identify a subset of the set of possible individuals. Another suggestion is that something is an individual if and only if the word that denotes it in the language is *impredicable*. This view was put forward in Aristotle's *De Interpretatione*, and re-emerged in the Middle Ages via the commentary on Aristotle's work written by Boethius. It amounts to the claim that something is an individual if and only if the word(s) that we use to refer to it can *only* appear as the subject of a phrase. Many medieval philosophers maintained that this is a satisfactory definition of individuality. However, this is mistaken. Ramsey [1925] argued that being subject or predicate at the level of language is a relative notion, for we can always reformulate our expressions in such a way that what appears as a predicate in one expression appears as the subject in another, equivalent one, and vice versa. More generally, the ontological difference between (i) something located in space and time that is attributed a quality that it shares with other things, and (ii) something that is exemplified by specific particulars and can be common to many things in different places and at different times, does not seem to be fully captured by a definition *exclusively* based on features of language. Another definition is proposed by Gracia, who emphasises the notion of *non-instantiability* and contends that:

> There is no great advantage in making a distinction between particularity and individuality [...because...] unlike singularity, which has plurality as its own opposite, particularity has no appropriate opposite of its own context, allowing us to use it as a synonym of 'individuality', and to oppose it to universality. [1988; 53]

This view is in itself consistent, for once individuality is *equated with* particularity no contradiction arises. However, there are reasons for *not* equating the two concepts. In particular, it seems possible, and indeed advisable, to take particularity as defining the general category of non-instantiables; and then distinguish – within the class of particulars – between individuals and non-individuals, depending on whether or not they have both determinate self-identity and numerical distinctness (this will become relevant in later sections of this chapter). Therefore, non-instantiability will not do for our present purposes, as it is necessary, but not sufficient, for individuation.[2]

The only option remaining is, thus, the Leibnizian view according to which every individual is qualitatively unique, that is, differs from other things with respect to at least one qualitative aspect.[3] Indeed, not only is this view not as easily refuted as those considered a moment ago, but it has, in fact, become a very authoritative view in metaphysics – and probably the dominant one among scientifically-minded philosophers.

A clear, and certainly very significant, opposition thus emerges that deserves careful examination. It is the opposition between:

a) REDUCTIONISM. The view that the world is, at root, entirely constituted by qualitative facts (i.e., facts other than those concerning identity and number of specific things), and individuality is consequently reducible to properties;

b) PRIMITIVISM. The view that the individuality of things is something over and above the qualitative aspects of those things as there can be primitive, ungrounded metaphysical facts of self-identity and numerical distinctness.

In the terminology introduced by Adams [1979], the former approach takes the things' *suchnesses* as the only relevant factors, while the latter maintains that some form of *thisness* also exists.

Reductionism was upheld, as already mentioned, by Leibniz. In its contemporary formulation, it is importantly related to Quine's (e.g., Quine [1960; 230]) idea that the identity relation can be treated as something that is not a linguistic primitive. The Leibniz–Quine reductionist perspective can be summarised as the view that the Principle of the Identity of the Indiscernibles (PII from now onward) holds, according to which (P being a property and x and y two property-bearing entities):

$$\forall x\, \forall y(\forall P(Px \leftrightarrow Py)) \rightarrow (x = y)).$$

Literally, PII says that if two entities have all the same properties, then they are the same entity. This is taken to ground individuality in the sense that each individual has a set of properties unique to it, i.e., that individuality is the same as qualitative uniqueness.[4] Whether or not one believes that this is actually the case, i.e., that PII is a truly compelling principle, largely depends on which properties the third universal quantifier in the above formula ranges over. Indeed, if properties such as 'is identical to a' or 'is numerically distinct from b' are included, then PII is analytically true. However, it is a widespread (and well-motivated) opinion that, at least within the reductionist camp, regarding numerical identity- and difference-involving properties as being on a par with all the others is illegitimate (i.e., that only qualitative properties should be considered). Indeed, doing otherwise would mean that identity is presupposed rather than analysed, with which reductionism would be based on an unacceptable circularity, and the very opposition between primitivism and reductionism lost.

Even setting this aside, however, different possibilities are available. Leibniz committed himself to a strong version of PII excluding spatial location and all relational properties from the scope of the universal quantifier ranging over properties. Nowadays, however, this restriction appears both unmotivated and such that it leads to a false principle. (Think of clones, or simply of Newtonian point particles.) A first question is, then, whether a PII ranging over all monadic intrinsic and relational properties (including location) can be plausibly regarded as true, given the evidence available to us. In classical mechanics (CM), a presupposition of impenetrability is generally made. In the third *regula philosophandi* of book III of the *Principia* [1687(1999)], Newton included in a list of fundamental properties of matter (together with, e.g., hardness, capacity of motion, and inertia) *impenetrability*, the property determining that each body always has a unique location in space. This, however, was neither an axiom of the theory nor a dispensable assumption, as it plays a fundamental role in CM. Usually, it is simply taken for granted: for, it is normally postulated that only force functions that satisfy certain continuity assumptions are allowed in CM, and these lead to equations that have unique solutions – so also guaranteeing uniqueness of location in space. With this proviso in mind, in what follows classical entities will be said to always differ at least with respect to space–time location. As a consequence, PII as defined above seems to be necessarily true in CM.[5]

Impenetrability is not, however, a (quasi-)axiom in the theory that we now take as the correct (non-relativistic) description of the fundamental constituents of reality, namely, quantum mechanics (QM). Let us see why, in some detail.[6] First of all, something must be said about quantum properties. The formalism of QM is based on Hilbert space, a particular generalisation of Euclidean space with an infinite number of dimensions. In such space, physical systems are represented by vectors, and properties by *Hermitian*[7] operators (i.e., specific functions that apply within the vectorial space). The eigenvectors[8] of each operator represent the possible values of the observable quantity represented by that operator. Now, given a physical system in state Ψ, an observable A[9] and a possible value v_i for that observable, the inner product[10] $\langle \Psi | P^A_i | \Psi \rangle$ – with P^A_i being the operator that 'projects' the vector corresponding to Ψ onto a ray[11] containing v_i as an eigenvector for the relevant observable – establishes a relation between Ψ and v_i. And if Ψ and v_i coincide, which naturally expresses the fact that the state has the value for A represented by the eigenvector v_i (it simply *is* the state in which the observable has that value), then the inner product is equal to 1 (for the vectors under consideration are all 'normalised', i.e., have unitary length, and $\cos(0°)=1$); while if Ψ and v_i are orthogonal the inner product is equal to 0 (as $\cos(90°)=0$), which can be straightforwardly taken to mean that the state does *not* have the value in question. Values in the entire range from 0 to 1 are also possible. In view of this, the theory lends itself naturally to an interpretation in terms of *probability assignments*, expressing how likely it is that, *upon measurement*, a physical system will turn out to have a certain property. Indeed, the statistical algorithm known as the *Born Rule* plays a crucial role in QM, and it establishes a connection between inner products in Hilbert space and probabilities as follows:

$$\mathrm{Prob}(o_i)_0^{|\Psi>} = <\Psi \mid P^o_i \mid \Psi>.$$

(This gives the probability that a measurement of the observable corresponding to the operator O on a system in state Ψ yields result o_i).[12]

What is important for present purposes is that, on the basis of these premises, it can be shown that whenever they are part of the same physical system, particles that share all their state-independent properties (i.e., roughly, particles of the same kind, having all the same essential properties) such as mass or charge – called *identical* particles by physicists[13] – also share all their monadic and relational state-dependent (i.e., accidental) properties, *including spatial location*. In this case, particles are said to be *indistinguishable*, and they are attributed exactly the same probabilities for measurement outcomes involving their properties.

This was first explicitly shown by French and Redhead [1988]. They considered two-particle systems of identical particles and an observable O with eigenvalues x and y; and analysed both *monadic* properties of the form Prob $(x)_{oi}^{|\Psi>}$ – that is, those expressed by the probability that in the state Ψ observable O 'actualises' upon measurement with value x for particle i; and *relational* properties of the form Prob $((x)_{o_1}|(y)_{o_2})^{|\Psi>}$ – that is, corresponding to the conditional probabilities of one value being actualised for O in one particle, *conditional on* the actualisation of the other value for the same observable in the other particle.[14] Deriving values for these probabilities from the quantum formalism, French and Redhead concluded that, both for fermions and for bosons,[15] two indistinguishable particles have the same monadic properties and the same relational properties one to another [Ib.; 241]. To see exactly why, consider the following.

We can introduce a *permutation operator* $Perm_{1,2}$ having the following properties. First of all, $Perm_{1,2}|a>_1|b>_2=|b>_1|a>_2$ for any $|a>_1|b>_2$ (that is, $|ab>$) representing the system constituted by a and b in the Hilbert space which is the tensor product of the two separate Hilbert spaces for a and b. From this, it follows that $(Perm_{1,2})^2=I$ (with I being the identity operator), and so $Perm_{1,2}=Perm_{1,2}^{-1}$.[16] Moreover, the operator has the same characteristics as those representing 'proper' observables, and this entails that it is its own adjoint,[17] and so $Perm_{1,2}^\dagger=Perm_{1,2}$. Hence, the permutation operator is unitary, and $Perm_{1,2}^\dagger=Perm_{1,2}^{-1}$. Therefore, one obtains that $Perm_{1,2}^\dagger Perm_{1,2}=I=Perm_{1,2}Perm_{1,2}^\dagger$. Now, the permutation operator acts as a unitary transformation of any operator O. Suppose O is considered with respect to two systems, namely, that $O_{12}=O_1\otimes O_2$. One has that $Perm_{1,2}^\dagger O_1\otimes I_2 Perm_{1,2}=I_1\otimes O_2$ and that $Perm_{1,2}^\dagger I_1\otimes O_2 Perm_{1,2}=O_1\otimes I_2$. Therefore, $Perm_{1,2}^\dagger O_{12}Perm_{1,2}=O_{21}$. Additionally, for any (anti-)symmetric quantum state the *Indistinguishability Postulate* (also known as *Permutation Invariance* or *Permutation Symmetry*) holds, according to which for any *n*-particle state and observable O on the *n*-fold tensor product state space, $<Perm\ \Psi|O|Perm\ \Psi>=<\Psi|O|\Psi>$ (with *Perm* being the operator associated with an arbitrary exchange of particles). Given the above, for any observable O and value x for that observable one obtains that:

$$Pr\,ob(x)_{o_1}^{|\Psi>}=<\Psi|P^{O1}_x|\Psi>=<\Psi|Perm_{1,2}^\dagger P^{O2}_x Perm_{1,2}|\Psi>$$
$$=<Perm_{1,2}\Psi|P^{O2}_x|Perm_{1,2}\Psi>=<\Psi|P^{O2}_x|\Psi>=Prob(x)_{o_2}^{|\Psi>}.$$

Since the choice of observable and value has been left absolutely unspecified, the above result allows one to conclude generally that identical particles in the same physical system have all the same monadic properties. As for relational properties, one can prove that $Prob((x)_{o_1}|(y)_{o_2})^{|\Psi>}$ $=Prob((x)_{o_2}|(y)_{o_1})^{|\Psi>}$ as follows. By a fundamental property of probabilities,[18] the above equality is the same as

$$Pr\,ob((x)_{o_1}\&(y)_{o_2})^{|\Psi>}/Pr\,ob(y)_{o_2}^{|\Psi>}=Pr\,ob((x)_{o_2}\&(y)_{o_1})^{|\Psi>}/Pr\,ob(y)_{o_1}^{|\Psi>}.$$

The denominators of this expression have just been shown to be equal. But the numerators are also equal. For,

$$\text{Pr}\,ob((x)_{0_1} \& (y)_{0_2})^{|\Psi>} = <\Psi\,|\,P^{01}{}_x P^{02}{}_y\,|\,\Psi> = <\Psi\,|\,\text{Perm}_{1,2}{}^{\dagger} P^{02}{}_x (\text{Perm}_{1,2})^2$$
$$P^{01}{}_y\,\text{Perm}_{1,2}\,|\,\Psi> = <\text{Perm}_{1,2}\Psi\,|\,P^{02}{}_x P^{01}{}_y\,|\,\text{Perm}_{1,2}\Psi> = <\Psi\,|\,P^{02}{}_x P^{01}{}_y$$
$$|\,\Psi> = \text{Prob}((x)_{0_2} \& (y)_{0_1})\,|\,\Psi>.$$

From this, it follows that *all* state-dependent properties are the same for indistinguishable particles in the same system.[19] The upshot is that indiscernibility seems to be an actual feature of quantum entities, clearly mirrored in the formalism.[20] Before starting to discuss the philosophical consequences of this, further considerations need to be brought to bear.

The *Exclusion Principle* (EP) has sometimes been referred to (for example, by Weyl [1927(1949)]) as a vindication of PII for *fermions*. Since EP bans two indistinguishable fermions from having all the same quantum numbers,[21] it seems to entail their discernibility. However, as first pointed out by Margenau [1944], EP notwithstanding, fermions in the same physical system do indeed have the same values for *all* their observables (provided, of course, that properties are identified with pre-measurement probabilities as assumed so far).[22] The result just arrived at, then, seems to hold. On the other hand, it is a fact that identical fermions in the same system have all the same properties, but we also *know with certainty* that, starting from a condition of *entanglement*[23] of the sort already discussed in passing in the previous chapter (when relating constructive naturalism and so-called 'experimental metaphysics' via a discussion of EPR-Bell scenarios), they will give rise to opposite results when measured. And this does seem to point towards an actual fact of the matter concerning some sort of qualitative difference.

Saunders has recently ([2003], [2006]) argued that this is indeed the case, and the difference being pointed to is due to ontologically fundamental *relations* – neglecting which without justification led to the impasse just described. If one assumes that relations exist that are not reducible to monadic properties, and has PII quantify over these relations too, then a weaker form of the principle than those considered so far emerges, which can be rendered as follows (R standing for a relation):

$$\forall x\,\forall y[[\forall P(Px \leftrightarrow Py) \wedge \forall R(R(x,y) \rightarrow \exists z(R(z,z)))] \rightarrow (x=y)].$$

And since participation in an irreflexive relation is sufficient for rendering the antecedent of the relevant conditional false, it also suffices for numerical distinctness, hence individuality. This form of discernibility has come to be universally defined 'weak discernibility' in the recent literature.[24]

Weak discernibility is evidently relevant for a reassessment of what has become the *locus classicus* with respect to the philosophical significance of PII, namely, Black [1952]. Black constructed a thought experiment involving a completely symmetric universe in which there are two numerically distinct spheres having all the same monadic properties and nothing else. There, whatever can be predicated of one sphere which is not an intrinsic property is necessarily a property the sphere possesses in relation to the 'other' sphere; but if any such relation must be expressed – as required in order to avoid trivialisation – completely in descriptive terms, then the spheres' relational properties are also equal [Ib.; 156]. In particular, spatial position, says Black [Ib.; 157–158], must be defined in relational terms because only the two spheres exist, and no absolute space–time has been posited. But both spheres have the relational property of being a certain distance away from a sphere with such-and-such properties. In this hypothetical universe, PII appears to be violated, as that one sphere is distinct from the other sphere seems to be a primitive fact not grounded in qualitative differences. Still, if one has recourse to weak discernibility, the two spheres might be said to be discernible after all, for being at a (non-zero) distance from something is an irreflexive relation that can only hold if there is more than just one object.[25]

Getting back to QM, Saunders' claim is exactly that fermions in the singlet state of spin[26] are weakly discernible. In particular, he argues, they are discernible because they are in an irreflexive relation expressed by the predicate ' ... has opposite ↑-spin component of spin to ... ' [2006; 59]. As a matter of fact, such relation is exhibited by the *total system* with probability 1. And given the (left-to-right part of the) eigenstate-eigenvalue link (EEL) – stating that a quantum system can be said to actually possess a given property *if and only if* the theory tells us that it will exhibit it upon measurement with probability 1 – this means that the relation exists with certainty, *independently* of the fact that the entangled fermions have the same probability assignments for their spin values, and in spite of the fact that no determinate spin properties can be attributed to the separate particles. In more detail, entangled fermions are not in *pure* states,[27] but only in *mixed* states.[28]

For instance, consider two entangled fermions, labelled 1 and 2. F for fermion 1, one has that its spin state in the relevant direction is:

$$1/2(|{\uparrow}{>}_1{<}{\uparrow}| + |{\downarrow}{>}_1{<}{\downarrow}|).$$

Analogously for fermion 2. But the composite system is in a pure state:

$$1/\sqrt{2}(|{\uparrow}{>}_1|{\downarrow}{>}_2 - |{\downarrow}{>}_1|{\uparrow}{>}_2)$$

Recalling the relevant statistical algorithm and noticing that[29]

$$< \Psi \mid P^o_i \mid \Psi > = |c_i|^2$$

and so

$$\mathrm{Prob}(o_i)_0^{|\Psi>} = |c_i|^2$$

one obtains that (with S denoting the observable corresponding to the chosen component of spin):

$$\mathrm{Prob}\,({\uparrow}_1|{\downarrow}_2)_S^{|\Psi>} = \mathrm{Prob}\,({\downarrow}_1|{\uparrow}_2)_S^{|\Psi>} = 1/2.$$

This entails that, *necessarily*, the component particles *have* opposite spin values in that direction (and that the total system – given EEL – *has* null spin in the relevant direction). This in turn means that there is a relation R holding between any two fermions a and b in the singlet state and such that Rab, Rba, $\neg Raa$ and $\neg Rbb$,[30] which means that quantum fermions are, indeed, (weakly) discernible after all.

Saunders' argument has been made more general in Muller and Saunders [2008] and Muller and Seevinck [2009]. The more general argument (from now on, the 'MSS argument'), it is important to note, is independent of probabilities, measurements and the very notion of collapse of the wavefunction – hence, from the assumption that the correct description of the quantum domain is that provided by the so-called 'standard', or 'orthodox' interpretation. In particular, it is emphasised that the discerning correlations are *categorical* (i.e., non-probabilistic and non-dispositional). What MSS do is to consider first an eigenbasis belonging to an *arbitrary* (physically meaningful) operator and the associated 1-dimensional projectors. They then define a single-particle projector corresponding to the *difference* between two such 1-dimensional projectors; that is, if P_m is the 1-dimensional projector on the ray to which eigenstate $|\Phi_m\rangle$ belongs, P_{lm} will be equal to P_l-P_m and will be such that

$$P_{lm}^{(1)} = P_{lm} \otimes 1; \quad P_{lm}^{(2)} = 1 \otimes P_{lm}.$$

The next step is to define relations on the basis of the sum over all possible eigenvalues,

$$R_t(a,b) \quad \text{iff} \quad \sum_{l,m=1}^{d} P_{lm}^{(a)} P_{lm}^{(b)} W = tW$$

and show that at least for some observables the 'composite projectors' produce eigenvalues *that can only be possessed by the system if the latter is constituted by more than one particle* (so making the corresponding relations satisfy the requirements for weak discernibility). For an example additional to that considered by Saunders, a pair of identical fermions in the singlet state of spin is necessarily in an eigenstate for the Pauli spin operator (equal to the difference between the + and − spin projectors for a given spin component of a single particle). The above can be generalised to all particles (including bosons) and Hilbert spaces of all dimensions. In particular, only spin degrees of freedom (specifically, total spin relations) need to be used to discern in infinite Hilbert spaces; and more general commutator relations holding between distinct single-particle operators (e.g., position and momentum) do the required work in the case of infinite Hilbert spaces.

Have we reached the conclusion that reductionism is correct in the quantum case as it was in the classical case? If so, should we be reductionists about individuality?

1.1 A first evaluation

It must be noted, first of all, that MSS distinguish *individuals* (which they take to be objects that are discernible on the basis of monadic, intrinsic or relational, properties) and *relationals* (according to them, merely weakly discernible objects), and consequently classify quantum particles as relationals, and *not* as individuals. While this may seem relevant for the opposition we are working with here (between reductionism and primitivism), it is not: for, MSS just work with a different, more restrictive, concept of individuality (essentially, one that presupposes a Leibnizian reductionist framework and equates individuality and absolute discernibility, i.e., difference in monadic properties) and, because of this, have to make room for an additional, *sui generis* category of entities. Of course, this different definition has a precise a theoretical basis. But identity and if one sticks (at least for the moment) with the general definition of individuality proposed at the beginning of this chapter there is no need to talk about relationals, and the results discussed in this section can be taken to support reductionism. Given the above, indeed it looks as though reductionism can be preserved – in spite of the evidence coming from quantum mechanics – if (and only if!) one accepts the possibility of merely weak discernibility. This immediately raises the question of whether there is anything objectionable in the above reconstruction of the arguments for weak discernibility in quantum mechanics.

One initial, purely philosophical, interrogative is whether relations can play the role the arguments in question require them to play – first and foremost, by being ontologically prior to or, at least, on a par with their relata (the identity of which they *determine*).[31] Indeed, whether relata must 'come first' is object of debate at least since the work of Russell [1911], and is definitely a complex issue.[32] However, the metaphysical possibility that not all relations are dependent on, and perhaps entirely reducible to, objects cannot be ruled out *a priori*, certainly not on the basis of mere intuition and 'linguistic practice' (i.e., because in logic relations are defined on the basis of n-tuples of individual objects). Hence, we will not spend more time on this alleged difficulty. Rather, assuming an appropriate naturalistic stance, we will look at the way in which physics lends support (or fails to lend support) to the proposed metaphysical framework based on this conjecture, and at the costs and benefits entailed by the choice of actually endorsing such framework.

One pressing question in this sense is, of course, whether the relations constructed by MSS truly are as physically meaningful as the monadic properties they are constructed from. MSS seem to think that they are, because they are built on the basis of uncontroversial single-particle operators. But consider the following. Muller and Seevinck themselves explain that their proof leads one to take two identical bosons in a symmetric direct product state as discernible, while the state in question is one in which two *completely similar, but non-interacting*, entities are simply considered together. They insist that this can be taken as a brute fact, much like entanglement. However, the evidence just considered might equally well be used to question the soundness of the MSS argument. That is, the result that independent bosons in identical states appear to be weakly discernible could be readily explained by claiming that the correlations presented as discerning are nothing more than *physically empty formal constructions* out of (admittedly, physically meaningful) single-particle observables. But, of course, if room is made for doubt in one specific case, it can be suggested that the claim that the constructed relations are not (at least, not necessarily) physically meaningful and metaphysically genuine gains credibility more generally.[33]

A different line of criticism (Dieks and Versteegh [2008], Ladyman and Bigaj [2010]) is that weak discernibility does not vindicate PII, because it does not correspond to the possibility of actually telling particles apart from each other through physical means and is therefore completely extraneous to the spirit of the Leibnizian tradition. (This is also relevant with respect to MSS' differentiation, mentioned above, between individuals and relationals.) Of course, MSS need not be impressed by this sort of arguments, as they can insist on the physical basis of their claims and the peculiarity of weak discernibility. However, the fact remains that since weakly discernible entities cannot just be picked out by physical means, the question emerges again concerning the ontological status of the allegedly discerning physical relations, and we are led back to the previous difficulty.[34]

The real problem, at any rate, is another. Even assuming that relations can obtain that determine, rather than being derivative on, the existence/identity of their relata as numerically distinct entities, and that there are genuine such relations in the quantum scenario, such relations are constructed in the case at hand on the basis of single-particle observables. But, necessarily, this means to *rely* on *already established* facts about numerical distinctness! If such facts were not already given, since one would not yet be in a position to say what individual objects exist, one would likewise not be in a position to determine which constructed relations have an actual physical counterpart in virtue of being constructed out of genuine monadic properties. It is crucial to understand exactly in what sense this is to be considered relevant here. MSS are well aware of the fact just pointed out, and in response to a common criticism levelled on the basis of it[35] contend that there is in fact no *circularity* here (between facts of distinctness and facts of discernibility). This is correct: all MSS' argument requires is that countability be assumed at the *formal* level, while the relevant conclusions concern the *concrete*, physical domain. Indeed, MSS argue, that the relevant *language* includes certain individual constants does not by any means suffice for identifying physically meaningful properties involving objects that those constants are taken to refer to. As a matter of fact, since the latter properties need to convey more than just information about identity and distinctness, physical discernibility is far from warranted by the assumption of countability at the formal level (indeed, it is patent that MSS do more than just state an obvious fact). This is convincing. However (and this is the key point, circularity objections notwithstanding), one is left wondering whether countability truly is a merely formal matter. That is, whether *formal countability does not have a direct counterpart in the actual world* in the form of basic facts of identity and distinctness that in fact *ground*, rather than being dependent on, the qualitative facts that MSS put so much emphasis on. In a word, could QM not be taken to suggest that countability is in fact metaphysically prior to discernibility? If so, reductionism as defined at the beginning of this chapter would obviously be in danger, as physical theory would lend direct support to primitivism.[36]

How should one react to the foregoing considerations? Should priority be given to reductive analyses of individuality or to putatively primitive identity facts? And why? Clearly, questions such as these call for a critical evaluation based on well-defined criteria of the sort indicated in the previous chapter. Before saying anything in that direction, though, another set of considerations must be brought to bear, having to do with quantum statistics.

1.2 Quantum statistics

Apparently, another blow to the entrenched belief that the world is ultimately constituted by tiny individual objects (more or less) like those we experience on a daily basis comes from a consideration of the statistical behaviour of quantum systems. Quantum statistics is, indeed, radically different from the classical statistics that we use to describe the behaviour of what we regard as paradigmatic individual objects. In CM, Maxwell-Boltzmann statistics (MB) holds. According to it, the number n_j of material particles in energy state j – given energy states in thermal equilibrium – is given by:[37]

$$n_j = g_j e^{-(\varepsilon_j - \mu)/kT}.$$

In the case of bosons, Bose-Einstein statistics (BE) applies. It has it that:

$$n_j = g_j/e^{(\varepsilon_j - \mu)/kT} - 1.$$

In the case of fermions, instead, one has:

$$n_j \; 5 \; g_j/e^{(\varepsilon_j - \mu)/kT} \; 1 \; 1.$$

The latter expresses Fermi-Dirac statistics (FD).[38]

Generalising to all observables, and focusing on the number of possible arrangements rather than on the number of particles in a given state, one has three different ways of counting the number of *a priori* equiprobable states in which physical systems could be found to be. Suppose one has N particles distributed over M possible single-particle microstates. In classical mechanics (with distinguishable particles),[39] the number of possible distributions W is

$$W = M^N$$

This is not true in the case of quantum particles, for which a smaller number of arrangements is available. For bosons, one has

$$W = (N + M - 1)! / N!(M - 1)!$$

and in the case of fermions, EP applies and further reduces the number of possible states, that becomes equal to

$$W = M! / N!(M - N)!$$

On the basis of these equations, one can calculate the probability for a specific configuration being realised, which is given by

$$Prob(s) = T / W$$

with *s* being the arrangement in question, and T the number of ways in which *s* can be realised (obviously, to be calculated via the type of statistics appropriate for the sort of entities being dealt with). Remarkable differences then emerge among the three statistics that can be illustrated through very simple examples, such as the following: classically one has four possible arrangements for every macrostate composed of two individuals to each one of which two states are available and equiprobable (e.g., two fair coins), and each arrangement has probability 1/4; considering the quantum-mechanical analogue of this scenario, instead, there are only either three such arrangements (for bosons) or one (for fermions) – and the probabilities are 1/3 and 1, respectively. The key difference is that permutations of qualitatively identical particles lead to statistically distinct configurations in the classical case *but not in the quantum case*. In particular, classical systems can be in *non-symmetric states* (that is, states in which individuals have definite but different values separately and for which, consequently, permutations do make a difference), while quantum systems cannot (but can instead be, as we have seen, in entangled states – which are also (anti-)symmetric). Using the customary notation, and considering again a two-particle system and a (generic) two-valued observable, the available possibilities can be represented as follows (with x and y representing the available values for the observable, and the subscripts indicating the – alleged – particle identities):

$$| x >_1 | x >_2 \qquad\qquad\qquad (C1-Q1)$$

$$| y >_1 | y >_2 \qquad\qquad\qquad (C2-Q2)$$

$$| x >_1 | y >_2 \qquad\qquad\qquad (C3)$$

$$| y >_1 | x >_2 \qquad\qquad\qquad (C4)$$

$$1/\sqrt{2}(| x >_1 | y >_2 \pm | y >_1 | x >_2) \qquad\qquad\qquad (Q3)$$

C1–C4 are the states available in CM, Q1–Q3 those available in QM (in particular, for reasons already given, Q1, Q2 and Q3 – with a negative sign – are accessible states for bosons, while only Q3 – with a positive sign – is a possible state for fermions). This is obviously very relevant from the philosophical perspective. Even more so, indeed, than the seeming failure of (traditional versions of) PII. For, it now seems that there is no space for individual objects in the quantum domain *even independently of whether their individuality is reducible to qualitative uniqueness or not*. So, what can be said about the evidence just illustrated?

The traditional explanation of the peculiarities of quantum statistics, which some (for example, French and Krause [2006]) call the 'Received View', is that particles are not individual objects, and this is why we should not expect states to be sensitive to which object has which property: if an object is not an individual, the argument goes, it doesn't have a well-defined identity, distinct from that of other objects, and thus there are no permutations to be made in the first place (nor can two objects have definite, distinct and separate properties as required for a system to be in a non-symmetric state – it makes no sense to attribute a property in a two-particle system to *this* particle instead of *that* one). On this construal, then, quantum particles would be non-individuals in the sense of being countable but not such that identity-involving formulas concerning them are well-formed. The resulting metaphysical framework requires non-standard logics (e.g., the Schrödinger logic of Da Costa and Krause [1994]) and semantics[40] and was developed in detail in recent years (see, for instance, French and Krause [2006] and [2010]).

However, while such a metaphysical picture seems to account perfectly well for the evidence and also agree, as its supporters repeatedly contended, with a natural reading of quantum field theory as the true theory of the microscopic domain (in terms of what, for instance, Teller [1995] dubbed 'quanta'), things are less straightforward than this. First, what quantum field theory suggests, if anything, is that what we regard as particles truly are 'perturbations' in – or excitations of – physical fields, where such perturbations are to be understood as *properties* (of the fields) rather than *objects*. For instance, if we take quantum field theory seriously, we should talk of the 'two-bosonness' of a certain electromagnetic field rather than of the presence of two bosons as objects there.[41] But, if this is so, talk of quanta and non-individual objects appears misleading. Second, and crucially, the shift to this alternative framework really seems *presupposed*, rather than motivated, by the 'results' that supporters of the Received View present in the non-relativistic case. How else could a non-individual-based ontology be extracted from an object-based theory and the failure of a principle such as PII? Indeed, it must be emphasised that the natural reaction to the failure of PII is to deny numerical distinctness, and in no way does the principle by itself entitle one to distinguish between individuals and non-individuals. Of course, it could be pointed out that we *know* that quantum field theory is preferable (perhaps because 'closer to truth') to non-relativistic quantum mechanics. But notice that the question being asked presently concerns *exclusively* the best interpretation of non-relativistic quantum mechanics!

Setting this aside, assessing the plausibility of an explanation of quantum statistics based on non-individuality requires a careful consideration of several other elements. Before embarking in such an analysis, though, some alternative accounts must be looked at, which will also complete the sort of overview of metaphysical options begun in the previous section.

From the MSS perspective illustrated earlier, first of all, the claim would be that the statistics describes the correlations that make particles weakly discernible, hence numerically distinct, but these correlations do *not* establish the full-blown individuality of *this* particle as opposed to *that* one (this may be regarded as an additional source of support for the differentiation between individuals and relationals drawn by MSS). Since this explains quantum statistics without giving up identity, weak discernibility might thus appear preferable with respect to the non-individuality Received View. But there are other, even less revisionary, possibilities. One strategy for avoiding the Received View is closely connected with a particular stance that it is possible to take with respect to the so-called 'Gibbs' paradox'. The Gibbs paradox consists in the fact that (MB) incorrectly predicts that by mixing similar gases at the same pressure and temperature one experiences a change in entropy, and this requires the introduction of a N! factor excluding permutations in order to make entropy correctly extensive. (Doing this amounts to switching to what is known as 'correct Boltzmann counting'.) One available interpretation of this is that, at least in some cases, classical particles are as indistinguishable as quantum particles. Saunders [2006a], for instance, after rejecting the claim that classical indistinguishability is incoherent (Bach [1997], Van Kampen [1984]) argues that what we should infer from the evidence is that both classical and quantum particles are statistically indistinguishable. On this basis, Saunders provides an explanation for quantum statistics: the difference between classical and quantum statistics, he contends, is due to the breakdown, as one moves away from the macroscopic, classical realm and towards the microscopic, quantum one, of the approximation

$$(N_k + C_k - 1)!/(N_k!(C_k - 1)! \approx C_k{}^N{}_k / N_k!$$

(with C_k denoting, roughly, the number of states available to particles at a certain energy level, and N_k the particles in that region). Such a breakdown, says Saunders, is entirely determined by the fact that the equilibrium measure is continuous in classical phase space, and discrete in Hilbert space. This is to say, more or less, that while every point in configuration space is equally available to classical particles, quantum particles only occupy specific 'areas' and tend to 'group together' in ways that are directly mirrored by the statistics. Clearly, Saunders' argument makes no assumption whatsoever concerning particle identities. It follows that it constitutes an alternative to the Received View that preserves the intuition that reality is, at root, constituted by individual objects (not necessarily individual because weakly discernible: it looks as though the views endorsed by Saunders on PII and quantum statistics need not be bought as a package deal). There are at least two difficulties with this sort of argument, however. First, the assumption of classical indistinguishability remains controversial. Second, an ontological account of the nature and behaviour of what is being described, going beyond a mere claim that there is a difference in probability measures, may still be legitimately demanded. *Why* are the probability measures different in the classical and quantum case? Why, more importantly, are only (anti-)symmetric, permutation-invariant states allowed for quantum systems? Why do quantum particles tend to group together in the way they do?[42]

While Saunders' explanation of quantum statistics rests on the assumption that all particles, classical and quantum, are indistinguishable, the reverse possibility was explored by Belousek [2000]. Belousek argues that whether quantum systems truly are permutation invariant depends on whether it is correct to assume the Fundamental Postulate of Statistical Mechanics (FPSM), according to which every distinct equilibrium configuration must be assigned the same statistical weight in the framework of quantum mechanics. Such an assumption, Belousek claims, is by no means inescapable. In actual fact, he argues, quantum particles can legitimately be regarded as distinguishable and, consequently, as individuals. For, as shown by Tersoff and Bayer [1983], one can derive quantum statistics under a hypothesis of uniformly random *a priori* distribution of statistical weights over all possible microstates of the system, *including* states only differing by permutations. Just by assuming, in agreement with the basic axioms of probability, that each state is given a probability between 0 and 1, and that the sum over the probabilities for all states is 1, Tersoff and Bayer showed that the observed statistical distributions correspond to the average over these random probabilities. Therefore, Belousek points out, while *given* FPSM an assumption of distinguishability accounts for (MB) statistics and one of indistinguishability for quantum statistics, it is possible to obtain (BE) and (FD) statistics for *distinguishable quantum particles* by denying FPSM and postulating a random *a priori* distribution instead. In particular, if one gives up FPSM, one can claim that the same number of states is, in fact, available to classical and quantum many-particle systems (although, obviously enough, with different probabilities). FPSM is generally taken to hold because, in the absence of any specific information about the system, it seems natural to think that it could be in any of the available states with the same probability. FPSM is therefore rooted in what is known the Principle of Indifference. However, Belousek points out, the latter is itself object of philosophical debate, and far from obviously compelling. As a matter of fact, an assumption of random *a priori* probabilities may legitimately be regarded as logically weaker than one of equal *a priori* probabilities: why should Nature divide in absolutely equal parts, rather than just let things go as they may happen to? If this is correct, there is, indeed, room for abandoning FPSM and claiming that quantum particles are distinguishable individual objects. This line of argument has the advantage that it connects the defence of particle individuality with the application of the well-understood, and certainly intuitively plausible, concept of distinguishability to the allegedly fundamental material entities. However, there are problems concerning both its explanatory efficacy and its general plausibility. First, even if one accepts that (although this fact is 'masked' by randomness in the distribution of probabilities) particle exchanges do, in fact, give rise to new macrostates in the quantum domain, the problem that non-symmetric states are never observed remains. Moreover, as Teller and Redhead [2000] point out, once some information about the physical system being described is available, interference terms arise that make uniform priors necessary; for otherwise, they argue following Van Fraassen [1991; 417–418], one would have to assume definite pre-measurement values for the individual particles, which is in direct conflict with Bell-type no-go theorems (see previous chapter).

Some authors (for the first explicit statements of the view, see French and Redhead [1988] and French [1989]) suggest that in order to defend the individuality of quantum particles in spite of quantum statistics it is sufficient to postulate certain primitive and non-further-explicable state-accessibility restrictions. That is, that non-symmetric states are ruled out just because this is a fundamental feature of the world – in the same way as, say, the existence of quantum entangled states or the fact that fermions obey EP. Indeed, if the initial condition is imposed that the state is either symmetric or antisymmetric, and there are no transitions from a state of one kind to a state of the other kind, then only one of two possibilities is open to any quantum system, and these correspond to BE and FD statistics. The crucial question is, of course, whether the conjectured restrictions on what is possible and/or what is true at the level of initial conditions can be accepted as such. Some, like Huggett [1995], maintain that it is explanatory enough to claim that non-symmetric states are simply not in the (symmetrised) Hilbert space that correctly represents the actual world. One may object, however, that this line of reasoning refuses to seek explanations in cases in which it is legitimate to ask for them, and so the interpretation of quantum statistics based on the Received View should be preferred in virtue of the fact that it is more explanatory and less *ad hoc*.[43]

In view of the foregoing discussion, it seems legitimate to say that in the case of statistics too, as it happened with identity and discernibility, it is not clear what the metaphysical significance of quantum mechanics actually is. In particular, the traditional views on what counts as an individual object do appear to be in trouble. But no definite, 'positive' metaphysical indications emerge from a consideration of the relevant physics. The fundamental question is, of course, whether there is a way to stick to the idea that material reality is ultimately composed of individual objects without sacrificing other seemingly basic assumptions in order to obtain a non-classical statistics. More generally, is the peculiar evidence coming from the quantum domain to be interpreted in the sense that one should re-describe one's ontological interpretation of the theory altogether (as suggested, say, by the Received View)? Or can one update one's metaphysical assumptions so as to preserve an essentially 'traditional' ontology?

2. A conservative proposal

A first important question to be asked in the present context is the following: why should one feel compelled to be a reductionist about individuality in the first place – especially considering the fact that, as we have seen, physical reality may be compatible with such a metaphysical view, but certainly doesn't straightforwardly support it?

It seems to be a common thought that a plausible answer goes as follows. In philosophical debates about the ontological constitution of material objects, it is well known that PII is a necessary truth only if objects are 'bundles' of properties, and properties are universals, i.e., repeatable entities whose instances are numerically identical to each other.[44] In metaphysics, bundle views with universals have been endorsed on purely a priori grounds, having to do with the seeming inconsistency of the Aristotelian–Lockean concept of a bare particular exemplifying and unifying properties; and with the alleged plausibility of an account of similarity facts which reduces them to facts of numerical identity between property-instances.[45] As for the naturalist, he or she should believe PII and the reductionist thesis to be true because science is, at root, exclusively concerned with general, qualitative descriptions of reality. Since, that is, scientists are only interested in the qualities of things and have no use for postulated entities 'playing the same role' in the world, metaphysicians should consider qualitative facts fundamental – so effectively endorsing, although for peculiar, non-a priori reasons, the bundle theory and PII.[46]

Here, of course, we share the naturalistic approach. The point, however, is that it is not obvious that 'science is only about qualities' – or, if it is, that this entails reductionism or at any rate makes it clearly preferable. An essential fact to be pointed out is, first of all, that indiscernible objects do, or at least *can*, in fact make a qualitative, empirical difference! Consider the very simple case of a world with two exactly similar material objects: clearly, such a world exhibits twice the mass of a world with only one such object.[47] Indeed, it seems right to claim that the naturalist, aiming to only formulate scientifically-grounded metaphysical claims, need not, and should not, take indiscernibility as a criterion of individuation for *objects*. His or her focus should, rather, be on *worlds* (i.e., systems of objects, be they concrete or represented in models). Consider, for instance, Earman's claim:

> the PII...seems to me to lead to a plausible methodological rule for scientific theorizing: in choosing two theories, where one theory employs a descriptive apparatus that implies the existence of distinct possible states between which we cannot distinguish by means of an observation, while the other theory has no such implication, then, all other things being equal, the latter theory is to be preferred to the former. [1979; 268]

Substituting 'world' for 'theory' and 'qualitative arrangement' for 'state', this claim perfectly conveys our point and yet clearly expresses a compelling empiricist methodology. What should be discarded as superfluous is any unnecessarily complex description, but not all descriptions of worlds including indiscernible objects automatically count as unnecessarily complex. To be sure, *if* two worlds are absolutely identical qualitatively, and one contains two indiscernible objects while the other does not, *then* the latter should be preferred for reasons of conceptual simplicity and economy. For, only in the first world a question arises concerning 'which object is which' that is not relevant *at all* for an account of the experienced facts. However, the key role here is played by the first conditional, concerning worlds, not by the existence of indiscernibles alone.[48] Thus, it turns out that what is methodologically problematic from the naturalistic viewpoint is by no means the existence of indiscernible objects in the same world/system/model, but rather something else, concerning worlds/physical systems (and the corresponding models) directly.

Now, if – even after restricting one's attention to scientifically relevant properties – a (model of a) physical system constituted by n absolutely indiscernible objects is or may be physically (i.e., not only with respect to number facts) *different* from one with only one such object, important consequences obviously follow for our present discussion. For, once understood along the lines just illustrated, the requirement of 'empirical relevance' rightly set by naturalists on metaphysical speculation is met also by an ontological interpretation of QM that allows for distinct indiscernibles. That is, it does not in any obvious way lead one to embrace reductionism about individuality. As a matter of fact, since *countability* is (as, we have seen, MSS explicitly acknowledge) a fundamental fact about the quantum formalism, and one that plays a crucial role in arguments in favour of the weak discernibility of quantum particles, it becomes possible to take non-relativistic quantum mechanics to lend itself naturally to a primitivist interpretation. In particular, we argued in the previous chapter that, *all other things being equal*, one's metaphysics should be as little revisionary (i.e., as continuous with common sense) as possible. And here we have now seen that, given uncontroversial facts about the quantum realm, reductionism forces us either to follow the Received View and consequently revise our common sense ontology quite radically (at any rate, more than other features of QM such as wave/particle duality, the uncertainty principle, and entanglement already do) or to undertake complicate manoeuvres in order to defend a principle (PII) that appears to lack solid (scientifically respectable) foundations anyway. On the other hand, primitivism provides us with a much more straightforward account of the relevant domain without having the unpleasant consequences that it is too often alleged to have.[49]

This is not the end of the story, though, since we have not tackled the problem with quantum statistics yet. Moreover, one may argue that the failure of 'Leibnizian' reductionism does *not* entail the inevitability of primitivism. Ladyman, for instance, insisted in recent work (Ladyman [2007], Ladyman and Leitgeb [2008]) that identity facts may not be grounded on qualitative difference, but must in any case be *contextual*, i.e., extrinsic and determined by the system in which the object possessing it is 'inserted', although *not* analysable in terms of qualitative facts and properties.[50] This would mean that such facts would be primitive but not in the customary sense of being basic facts about particular objects independently of everything else. Remarkably, the reason Ladyman adduces for opting for this 'third way' between the traditional forms of primitivism and reductionism has to do, among other things, exactly with quantum statistics. In particular, Ladyman argues that if identity facts were not extrinsic, then the truth of 'haecceitism' (the doctrine that there can be differences between what distinct worlds say *de re* about certain individuals that do not correspond to overall qualitative differences – recall our distinction above between trans-world indiscernibility and indiscernibility as applied to objects) would follow; but, says Ladyman, haecceitism, besides being in conflict with naturalistic methodology, is contradicted by contemporary science. Consider, for instance, general relativity. There, absolutely indiscernible descriptions of space–time, only differing with respect to which point is which, should be regarded as identical if one is to avoid violations of determinism implied by the hole argument (Earman and Norton [1987]). This seems to mean that the points of the manifold – which do not seem to differ qualitatively[51] – do not have intrinsic primitive identities. Therefore, they must be contextually individuated by the non-qualitative relations that characterise the metric field as it is defined on the manifold.[52] Of course, the idea is that the same reasoning carries over to quantum statistics, where, as we have seen, permuting identical particles in many-particle systems does not give rise to a new, statistically relevant, physical scenario: one may legitimately think that the reason for this is that quantum mechanical many-particle systems of identical particles truly are, first and foremost, structures with a fixed number of 'places' that can be indifferently occupied by any one of the particles that play the role of 'place-holders' in those structures.

This may, at a first glance, appear to be a compelling argument, striking a perfect balance between explanatory power and revisionary nature of the proposed ontological framework (the relativity-quantum mechanics analogy has been questioned, but this need not detain us here). For, even insisting on countability facts and their being directly mirrored by the Hilbert space formalism, the traditional primitivist appears unable to explain the sort of permutation invariance exhibited by quantum as well as general relativistic systems of entities. How can such permutation invariance coexist with 'intrinsically grounded' identity facts? However, upon scrutiny, the contextualist arguments turn out to be less convincing.

First of all, primitive identity *need not* entail haecceitistic differences! For, what is true of distinct worlds is not univocally determined by the nature of the identity of the objects that inhabit them. Adams, for one, makes this point when he claims that the issue of

> whether the identity of the actual philosopher [Aristotle] with the possible tax collector ... is quite distinct from that of the qualitative or nonqualitative character of Aristotle's identity. [1979; 20]

A counterpart-theoretic treatment of possible worlds, for instance, allows one to assume primitive intra-world identities (i.e., primitive identity as the ground for the individuality of specific objects) together with anti-haecceitism.[53] Moreover, a supporter of the sort of 'super-essentialism' that many commentators attribute to Leibniz[54] may claim that, since individuals have all their properties essentially, no individual is identical to any other individual in any world. In addition to this, it is crucial to notice that, *even if haecceitism holds*, there might be (equally or more forceful) explanations of the evidence alternative to the contextualist view. That is, plausible ways of accounting for the fact that, even though permutations can in general give rise to genuinely new scenarios, this does not have empirically relevant consequences in the case at hand. Expanding on this, one option for the primitivist is thus to argue as follows. In classical mechanics, the assumption can be made that all the statistically relevant properties (i.e., state-dependent, non-essential properties) are possessed by individual particles as their monadic properties. This certainly holds in the domain of everyday objects, where it provides a straightforward explanation of why we expect, say, one of four possible outcomes in cases like that involving two fair coins. However, that this assumption holds is obviously not a metaphysical necessity, and a simple look at entangled states shows that, in fact, it is acknowledged not to always hold in QM – we have already seen in the previous chapter that entangled systems exhibit 'holistic' correlations that do not reduce to separate properties of sub-systems and are, in fact, independent of which subsystem is which and what property is exemplified by which subsystem. More generally, in the classical domain, Humean supervenience (the doctrine that the whole of reality can be reduced to local matters of fact about objects exemplifying monadic properties plus spatio-temporal relations – more on this in Chapter 5) holds; but this is not the case in the quantum domain.[55] Now, in light of this, it could be suggested that *all* state-dependent properties of quantum particles in many-particle systems (entangled *and* non-entangled) are holistic properties and are consequently independent of specific facts about specific particles and their monadic properties. From which it follows that particle exchanges do not, and cannot, give rise to new, empirically relevant states (i.e., new descriptions) – regardless of the truth or falsity of haecceitism. With this, one immediately obtains an account of the evidence that doesn't involve claims about the identity and individuality of things – i.e., at least from the perspective of the primitivist, a clear explanatory advantage. And notice that this also accounts for the impossibility of non-symmetric states: these latter states require separate, well-defined (and distinct) properties for the separate components of the total system, which is exactly what the present proposal rules out in all the relevant cases. The foregoing suggestion (articulated in more detail in Morganti [2009]) basically consists of an extension of the sort of holism outlined by Teller ([1986]) and Healey [1991] for entangled states to non-entangled states. However, it does not need to share the specific details of these views: most notably, with respect to Teller's proposal, it is not necessary here to assume that the relevant properties are 'non-Humean' *relations*. Indeed, if the holistic properties described by the quantum statistics of identical particles are regarded as *monadic properties* of the total system, the worry that 'empirically empty' haecceitistic permutations remain possible is conclusively dispelled – for, there no longer exist, say, two ways for two particles to be at the 'two ends' of an entanglement relation.[56]

Regardless of the exact nature of holistic quantum properties (again, more on this in Chapter 5), it might be objected that primitivism remains implausible from the naturalistic viewpoint because it posits the existence of primitive, intrinsic facts of identity that cannot but coincide with mysterious, purely metaphysical factors. That is, basically, with something like Duns Scotus' *haecceitates*, those additional ontological components that, according to Scotus, are required for 'contracting' common natures into full-blown individuals. But this is not a cogent objection. To begin with, what does it mean, exactly, for non-qualitatively grounded identity to be contextual? Ladyman suggests that identity and difference be included in the relations characterising the 'structure' to which objects belong (Ladyman [2007; 35]). But what is the metaphysical counterpart of these relations if not completely primitive *non-physical* entities? And what do these relations reduce to in the case of one-object systems if not to primitive ungrounded identities – albeit, perhaps, of places in structures? Primitivists and contextualists seem to be travelling in the same boat here. But rather than leading them to revert to Leibnizian reductionism, that boat can be a means to reach the conclusions that both positions are perfectly acceptable from the naturalistic perspective. To do so, one has to endorse a 'thin' form of primitive identity, intended as *not constituting a metaphysical addition* to the properties of things, and just coinciding with fundamental facts about the existence of certain entities as *those* entities. This would be a sort of 'Ockhamian', rather than 'Scotian', view, according to which

> [w]hatever is singular is singular through nothing added to it, but by itself. (Ockham, *Ordinatio* I.2.6, 86)

That is, individual objects have their unique identity primitively, without this meaning that there is an actual 'ontological component' of them that is responsible for this. In other words, the primitivist can subscribe to a form of nominalism according to which the fact of being a numerically unique object is a primitive fact, not something that is determined by an actual entity or factor that qualifies as a haecceitas or a primitive thisness.[57]

Summing up, we have found an account of quantum statistics that appears to fare well in terms of credibility and amount of revision required. It is credible because it simply extends to non-entangled states and their properties a treatment of entangled states and their properties that is commonly accepted as inevitable, given the evidence related to EPR-Bell scenarios; and, by so doing, it provides a straightforward account of *both* Permutation Symmetry and the impossibility of non-symmetric states. Compare this with contextualism, for instance: in embedding the quantum domain in a larger framework in which *all* identity facts are said to be contextually determined, it explains permutation symmetry but not the impossibility of non-symmetric states.[58] As for ontological revision, instead of taking the evidence as sufficient for re-describing our ontology in terms of non-individual *things* (Received View) and/or making the *identity* and *individuality* of things extrinsic (but not qualitatively analysable), the proposed account only requires a modification of our beliefs about the *properties* of things. And it seems sensible to claim that, in a hypothetical hierarchy defined on the basis of metaphysical conservativeness, the notions of object-hood and individuality are on a higher level with respect to the monadicity and non-relationality of all properties. (To repeat, the latter is something that current physics rules out explicitly anyway.)[59]

2.1 Final evaluation

The upshot of the foregoing discussion of identity and individuality in non-relativistic quantum mechanics is, thus, the following. There are clear indications coming from the physical theory that the classical concept of an individual object cannot be preserved in the quantum domain. However, the strong view that individuals must be left out of the picture altogether only follows if one takes discernibility with respect to monadic properties and classical statistical behaviour to be an essential component of the notion of individuality. It has been argued, though, that there are various alternatives, such as weak discernibility, classical indistinguishability and contextuality. Here, it has been additionally suggested that there are no compelling reasons (be they methodological or *a priori*) for the naturalist to endorse PII as a guide for the ascription of individuality to things; and that the peculiarities of quantum statistics can be accounted for without saying anything at all about identity facts. In particular, it is possible to regard quantum countability (as it is directly mirrored by the particle labels and names that are an integral part of the quantum formalism) as sufficient for endorsing primitivism; and to extend certain ideas that are normally applied to at least some many-particle quantum systems so as to make the peculiar features of quantum statistics depend on the property-structure of the domain it describes rather than on the identity conditions of the objects inhabiting that domain and exhibiting that property-structure.[60]

It should be already clear what this, looked at against the methodological background defined in the previous chapter, will be taken to mean here. Without undertaking a detailed examination of all the proposals that have been considered with respect to (in)discernibility and statistics, we will assume that, by and large, all of them fare well with respect to the 'traditional' theoretical virtues of accuracy, consistency, applicability, simplicity, and fruitfulness, as well as with respect to the criterion of refutability. With three important provisos, though:

1) The Received View, postulating the existence of particles as non-individuals, is not in fact as empirically *accurate* as one might wish, because it essentially ignores countability facts by switching to the Fock space formalism without names typical of a different theory (quantum field theory rather than non-relativistic quantum mechanics) – in fact *requiring*, we argued, an ontology of field-properties rather than non-individual objects in order to avoid an inconsistent use of PII.

2) Weak discernibility, as openly acknowledged by its supporters, is slightly less *simple* than the alternatives. For, not only does it accept the existence of non-supervenient, irreducible quantum relations – something quite uncontroversial given the evidence of the EPR-Bell type. It also postulates that, at least in some cases, relations can exist without depending (if not mutually) on their relata – clearly an ontological addition.

3) Contextualism provides a neat solution to the problem of permutation invariance, which also nicely extends to space–time points. But it is not obvious, as we have pointed out already, that, as it stands, it can explain the impossibility of non-symmetric states in quantum mechanics other than by simply ruling them out *a priori*. Which clearly means that the view is less *applicable* than alternatives that do provide such an explanation in a non-ad-hoc fashion.

Primitivism, instead, seems able to steer clear of these difficulties. At the same time, we have argued, contrary to a common opinion it is not less motivated than the alternatives from a scientifically-oriented perspective. Primitive facts of identity, at least when properly understood, do not automatically qualify as suspicious metaphysical hypotheses. Moreover, given that at the level of common sense the intuition seems to be that individual objects are such because of facts and properties intrinsic to them, i.e., in virtue of their essential metaphysical nature, of their 'way of being',[61] the primitivist account fares better with respect to the Quinean criterion of *conservativeness*. Certainly, it is less revisionary, at least in the sense that it does not require one to say anything about the ontological status of relations, nor about the possibility of objects that are not individuals.

3. Quantum fields and beyond

A natural reaction to the foregoing analysis is to ask whether it is enough to consider non-relativistic quantum mechanics, as we did so far, if we are to draw compelling metaphysical conclusions from our best physics. The answer to this question is certainly 'No'. For, it is part of the consensus in the physics community that theories other than non-relativistic QM should be regarded as providing a more appropriate description of the physical world, starting from quantum field theories. It must be immediately pointed out, though, that the priority of quantum field theory, and perhaps other theories, over traditional quantum mechanics does not mean that we have spent time in a useless discussion so far. For, first, non-relativistic quantum mechanics is still the most discussed theory in terms of philosophical consequences in the literature – if only because it is that on which the highest level of agreement has been found. Hence, the foregoing was useful at least in terms of clarification and critical assessment of an existing debate. Second, the interest in non-relativistic quantum mechanics is at least partly justified by the fact that any reflection on more sophisticated theories will be likely to use that theory as a starting point – even though it is likely to also involve aspects that were not contemplated in the non-relativistic case, too. Bearing in mind, then, that our case study in this chapter focused squarely, and legitimately, on non-relativistic quantum mechanics, let us more briefly look at what additional elements relevant for our discussion of individuality emerge when one moves beyond it.

3.1 Relativistic and non-relativistic quantum field theory

Let us begin with a very brief historical detour, so as to introduce the concept of a field and illustrate its physical status and significance.

For Newton, as for Gassendi and Boyle, particles could exert force (and perceive that of other particles) without being in direct contact with each other. The notion of interaction without direct contact was instead criticised by Descartes and Leibniz.[62] This latter insight was developed in the 19th century, when the likes of Young, Fresnel, Stokes and Thomson worked towards a thorough definition of the ether as a continuous medium constituted by dimensionless points through which light propagates. Against this background, it was then Faraday who modified the notion of a physical field from that of an *ad hoc* device for representing the actions of bodies to that of a full-blown physical entity. On Faraday's view, one could even dispense with material bodies altogether: instead of taking them as real things surrounded by fields, they can be regarded as by-products of the fields themselves. Such primacy of the ether over matter was also defended by Thomson (Lord Kelvin) and Larmor. With Einstein, of course, the ether was set aside. However, a realist view of fields can be formulated on the basis of Einstein's theories. This is illustrated by Lange [2002], for instance, who uses the example that – given the mutual connection existing in the relativistic setting between energy, momentum and mass – a nuclear decay results in the sum of the resultant bodies' masses being less than expected *unless one counts in a field element* (bearing the 'missing mass' in the form of energy). If one does so, Lange points out, one must be realist about the field.

The pioneers of quantum mechanics took particles to be the basic constituents of reality but soon extended their analysis to fields. Already in the late 1920s, Heisenberg, Born, Jordan and Dirac applied the methods of quantum mechanics to electromagnetism, replacing the classical variables taken to correspond to components of the field with quantum mechanical operators. More particularly, they conceived of the field's internal degrees of freedom as an *infinite* set of harmonic oscillators and applied a quantisation procedure to those oscillators. Important successes were obtained on this basis in the following two decades. In the 1950s and 1960s, the programme knew a period of decline:[63] in the early 1960s, for instance, Chew openly discarded the field approach in favour of an alternative programme, (the S-matrix approach). This crisis, however, was of short duration. By 1969, as Cao puts it,

> the quark-parton model, and the more sophisticated quantum chromodynamics within the field theoretic framework were accepted as the fundamental framework for hadron physics. [1997; 261]

Indeed, in the 1970s quantum fields re-emerged and rose to the status of fundamental entities in contemporary physics, a status that they still possess nowadays. An important cause for this is, of course, that quantum field theory is relativistically invariant in a way that is simply ruled out in the case of QM and thus, at least in some formulations, can be hoped to be the basis for what would be one of the most important steps forward in the history of physics: the unification of quantum theory and relativity (more on such unification in the next chapter).[64] But let us now get back to our topic in this chapter. What is the ontological nature of fields, exactly?[65]

First of all, it is essential to point out that, despite its name, quantum field theory (QFT) does not, by any means, have an obvious ontological interpretation, i.e., in terms of fields. For, even though its mathematical formalism describes fields, i.e., entities with an infinite number of degrees of freedom, the ontological counterpart of this is not straightforwardly determined. Indeed, as first explicitly pointed out by Teller [1990], the analogy between classical and quantum fields breaks down in at least one fundamental respect, namely to the extent that classical fields are such that there are definite physical magnitudes at all their space–time points, while in the quantum case one only has operators representing spectra of *possible* values, from which something physical *may* emerge at a later time. In general, field values attached to space–time points may be said to have no direct physical significance in the quantum case (Kuhlmann [2010]). This may encourage the conservative metaphysician to take QFT to really just be a 'refined' QM, and consequently insist on a particle-based ontology (although one that must make sense of the degrees of freedom possibly not being finite, which certainly constitutes a non-negligible element). Things, however, are not so straightforward.

First of all, 'creation-' and 'annihilation-events' play a crucial role in QFT, and this entails that (what would seem to be) particles cannot be attributed continuous paths across space and time (at least when the state of the system changes). For, every change of a physical state literally involves entities coming into and/or going out of existence. Another important fact is that it is possible, and indeed common, in QFT to encounter quantum states that are not eigenstates of the *number operator* (measuring the total number of particles present in the system). Such states cannot be handled unless they are described as quantum superpositions of states having different values of number. For example, suppose we have a bosonic field whose particles can be created or destroyed by interactions with a fermionic field. The Hamiltonian of the combined system would be given by the Hamiltonians of the free boson and free fermion fields, plus a 'potential energy' factor that includes an 'interaction term' describing processes in which fermions either absorb or emit bosons, thereby being kicked into a different eigenstate. In these cases, even if we start out with a fixed number of bosons, we end up with a superposition of states with different numbers of bosons. In condensed matter physics, states like these are particularly important: they are needed for describing so-called 'superfluids' (matter behaving as a fluid with no viscosity), as many of the defining characteristics of the latter arise from the quantum state being a superposition of states with different particle numbers. The concept of a 'coherent state', used to model lasers and the BCS ground state,[66] also refers to a state with a non-sharply-defined particle number. Things become even more complex when one brings relativity explicitly into play. For then, one must take into account:

1) No-go theorems on the localisability of particles, according to which relativistic QFT is unable to describe particles as localised in finite regions of space–time, for doing so would violate basic relativistic requirements as the impossibility of superluminal speed (Malament [1996], Halvorson and Clifton [2002]);[67]

2) The Reeh-Schlieder theorem (Reeh and Schlieder [1961], see also Redhead [1995]), which asserts that local measurements never allow one to distinguish a state with no particles ('vacuum state') from any n-particle state;[68]

3) The Unruh effect, consisting in the fact that a uniformly accelerated observer in a vacuum will detect a 'thermal bath' of particles (the so-called 'Rindler quanta'), so that – roughly put – a change in the frame of reference causes a change in the number of particles;

4) Haag's theorem (stemming from a conjecture by Haag [1955], for a thorough reconstruction and discussion see Earman and Fraser [2006]), entailing that in an interacting relativistic QFT, the customary representation of free particles cannot be also used to describe interactions;

5) The fact that expectation values for certain quantities do not vanish for the vacuum state, so that energy is not zero, and there are physical happenings even when no particles are there.

One way of summarising these results in terms of the number operator N is the following: (1) and (2) add to the problem of superposition mentioned earlier the fact that local number operators are not well-defined; (3) puts into question the objectivity of total number operators, (4) even questions their existence, at least in the interaction case; and (5) adds that there might be 'physical existence' even if $N=0$.

On the other hand, however, it has also been forcefully argued that opting instead for a field-based ontology for QFT is implausible. Baker [2009], for instance, regards the wavefunctional interpretation of QFT[69] as the most natural one and points out that 'wavefunctional space' is equivalent to many-particle Fock space, and thus the relevant results (especially those pertaining to (3) and (4)) carry over to the field-ontological setting. In view of this, the supporter of an ontology of individual objects might try to insist on his or her preferred view by making room for ontological revision. Indeed, since the term 'individual' is not synonymous to *'classical*-particle-like object', nor is the latter concept constitutive of the former, there is some space for manoeuvre here – exactly as there was in the case of QM (there, too, as we have seen, *some* presuppositions valid in the classical domain had to be dropped, but this did not *ipso facto* mean to abandon the notion of an individual object altogether). But where does this lead in the QFT case?

To begin with, Redhead [1982] suggested that in QFT particles are at least a logical possibility in the form of 'ephemerals', i.e.:

> entities which can be distinguished one from another at any given instant of time, but unlike continuants cannot be re-identified as the same entity in virtue of [primitive identity...] at different times. [1983; 88]

This may account for a number of things: the non-identity of the (putative) objects that may occupy the same space–time location before and after an annihilation/creation event; the fact that the total number operator is not fixed in QFT; and the non-localisability results à la Malament (1) above, which crucially rely on an assumption of fixed total number. Next, one may contend that individuals need not be sharply localisable (the assumption of localisability for particles is given up, for instance, by Saunders [1994]; and localisation has been made unsharp by Busch [1999], and hyperplane-dependent by Fleming and Butterfield [1999]); nor objective in the sense that the features of individual entities, perhaps even their existence, are completely frame-independent (see, again, the hyperplane-dependent approach championed by Fleming, e.g., Fleming [1989]). And this would represent a response to problems (1), (2) and (3) above. But this strategy seems to have an intrinsic limitation, determined by the last two results above, concerning the vacuum and the lack of a total number operator in the interacting picture. As for (4), one might try to neutralise it by pointing out that Haag's theorem is a problem for interacting QFTs in general, and/or that a well-defined total number operator is not needed for an ontology of individual particles. Bain [2000], for instance, suggested that a viable notion of particle can, in fact, be reconstructed for interacting fields by employing so-called 'LSZ scattering theory', where it is assumed that interacting systems tend to gain the status of free systems at asymptotic times (i.e., as the time variable tends to plus or minus infinity). Coming to (5), one can maintain that vacuum fluctuations are simple consequences of the time–energy uncertainty principle: any ordinary field cannot maintain zero value at all times, and virtual particle–antiparticle pairs of all types and all energy-momenta must instead continually form and annihilate at all points. Finally, it is perhaps worth mentioning also that, as to the possibility for the number operator N to be in superposition, one can sustain an ignorance interpretation of the relevant states (i.e., that there is an objective number of particles in the relevant systems, but we can only get to know it with certainty by measuring). There would, of course, be much more to say about these issues, and what was just said was merely intended to sketch some of the possible options available to the particle ontologist[70] – all of them definitely open to discussion. What is certain is that the costs entailed by the idea of preserving a traditional ontology of individual objects in spite of the evidence related to quantum fields are non-negligible. For instance, can one make do with number operators definable only at asymptotic times? Is it acceptable to make one's fundamental ontological posits – i.e., their number and even existence – frame-dependent? Can we be realists about virtual particles? Given the methodological criteria set out in the previous chapter – stating, among other things, that ontological conservativeness can be accepted only to the extent that our best (scientific) description of the relevant domain lends itself more or less naturally to the proposed interpretation – the foregoing appears sufficient for looking for alternatives. Besides, an ontology of individual objects now fares pretty badly not only in terms of applicability/fit with the available scientific data, but also with respect to the preservation of established beliefs based on well-defined concepts.

If this is correct, there seem to be two options remaining: either a form of scepticism/agnosticism about the possibility of defining an ontology for (relativistic) QFT; or the search for a revisionary ontological framework truly capable of making sense of the theory. Setting aside the former alternative as a live option which is, nonetheless, clearly a last resort for the naturalised metaphysician, we will close this section by offering a quick overview of the latter option.

Revisionary ontologies can, of course, be revisionary to various extents. Some possibilities that have been seriously considered in this sense are the following, which will be presented, and briefly commented on, in the (intuitive) order of their closeness to commonsense beliefs.

1. *Non-individuals.* Such an ontology was proposed by Teller with his abovementioned 'quanta' interpretation of QFT ([1990], [1995]) and, more recently, by French and Krause ([1995], [2006]). Indeed,

we acknowledged in the previous section that in the QFT scenario an ontology of non-individuals lacking well-defined identity might be justified. However, there are (at least) two things to point out: first, as we have seen, non-individuals are supposed to be at least cardinally countable, which they plainly are not in the relativistic quantum setting; secondly, as we have also seen, talk of quanta really seems to be best understood in terms of talk of properties of field-points (or regions) (see our earlier discussion of PII and non-individuality), but this suffices for switching to another ontology.

2. *Property-instances (tropes)*. One alternative is to focus on the *properties* of fields directly, bypassing the identification of systems, objects, things and fields possessing those properties. Connecting this strategy to the abandonment of the purely philosophical idea of properties as repeatable universals, this has led some authors to define an ontology of so-called 'tropes' (primitively individuated property-instances) for field theory. In view of the fact, reported above, that quantum fields do not normally exemplify categorical properties, i.e., properties that they possess with certainty and actually manifest, these might be intended as dispositional vacuum expectation values (Wayne [2008]); dispositional properties connected to representations of algebras of observables in algebraic QFT (Kuhlmann [2010]); or determinables assigned to smeared space–time regions (Lupher [2010]). Without discussing which one of these options is preferable (something that depends on one's metaphysical views as well as on one's ideas about algebraic versus canonical QFT), the problem with these attempts is that they are unable to solve all the relevant problems. To the extent that (as explicitly contended, for instance, by Kuhlmann) individual objects are supposed to emerge as real entities constituted by property-instances, the no-go results that apply to objects seem *ipso facto* to apply also to tropes. And even if one somehow excludes objects from the picture altogether, difficulties remain: for instance, it is still an open question, say, why the vacuum is not 'really empty'; or which, and how many, tropes exist at a given region independently of the frame of observation. Going fine-grained, then, does not help.

3a. *Events*. Event ontologies start from two observations: first, that in some cases identity and/or localisability criteria are simply not available for individuals but exist for events, in the form of the Davidsonian criterion of identification based upon causal location within a totality of causal connections; second, that events as the basic particulars of an ontology may fully satisfy all the requirements for a complete and consistent description of the material world in cases in which traditional individuals do not, especially when one switches to a relativistic setting – in which events are customarily taken as the basic ontological units. Based on considerations akin to these, Auyang [1995] takes point field-events as the basic items; Bartels [1999], instead, proposes to take events of the kind 'system S *enters/starts being in* (quantum) state Q' as the stuff which (quantum) reality is made of.

3b. *Facts/States of affairs*. A line of thought going back at least to Wittgenstein's *Tractatus* and taken up in David Armstrong's recent work (see, in particular, his [1997]) has it that reality is, at root, constituted by states of affairs, ways in which things are. For Armstrong, states of affairs are non-mereological composites of thin ('bare') particulars and properties [Ib.; 118] and must be regarded as basic. In the quantum context, this leads to an ontology close to that proposed by Bartels, where 'system S *is* in (quantum) state Q' becomes the paradigmatic ontological unit.

Now, even granting that events or facts are ontologically basic rather than dependent on objects and/or properties, events and facts too seem subject to at least some of the above no-go theorems. For instance, how do I identify and count objective events in a frame-independent manner? Are the different ways of determining the particle content of a given physical system – relevant for the Reeh-Schlieder theorem – not just different ways of counting events?

4a. *Processes*. Since the time of Heraclitus (6th–5th century B.C.), several philosophers regarded change as fundamental, and reality as having an essentially dynamic nature. This led to a viewpoint according to which processes encoding change and variation – rather than objects as invariances in these processes – are basic. Whitehead (see his [1929]) was the first to develop a complete process ontology, according to which processes actualise certain relations among those inhabiting a *sui generis* realm of possibilities. The Whiteheadian insight was revived by Stapp (e.g., in his [1979]) who focused on quantum mechanics and suggested that physical theory describes a global dynamical process; and by Seibt [2002], who defined an 'axiomatic process theory' in which four-dimensional non-countable entities able to exist in disconnected space–time regions, and possessing different degrees of determinateness ('free processes'), are fundamental (Seibt explicitly applies this ontological viewpoint to QFT).

4b. *Factors*. In a [2002] paper, Simons took as fundamental ontological items invariable universal constants which constrain the form of physical happenings. These 'basic factors' he conceives of as universal invariants identified on the basis of physical fundamental constants, and analogous to Empedocles' hot/cold and wet/dry pairs. In Simons' model, the basic factors are called 'modes' and come in 11 families dubbed 'modal dimensions'. The possible combinations of modes (3072) create all elements.

In this case, too, the problems for the proposed ontological accounts should not be too difficult to spot. Most importantly, the sort of entities being presented as fundamental seem hardly graspable unless they are analysed in more familiar terms, and physics does not seem to give us any reasons for taking anything like, say, Simons' factors as ontologically basic.[71]

In view of all this, without by any means presenting this as anything like a final pronouncement on the matter, let us suggest that perhaps, *given the current status of scientific research*, the sceptic approach might, after all, be the most sensible one when it comes to the ontology of QFT. For, it is not just that a lot of revision in our customary, commonsensical ways of seeing things is imposed by (relativistic) QFT when it is interpreted along the revisionary lines just reviewed – especially in some cases. It is also true that – again, as things stand now – no consistent, empirically adequate, fruitful and fully understandable ontological account of the theory has been suggested at all.

Perhaps, though, this conclusion could be avoided by having recourse to one last kind of revisionary ontology, based on structures. Such ontology deserves a separate, and more detailed, treatment for two reasons. First, because it has become very popular recently among philosophers of science and of physics. Second, because it is argued for not only on the basis of quantum field theory, but also of the facts about the quantum mechanical domain that we have already discussed – which might then come to be seen under a new light. We will look at this structuralist option in a moment. Before that let us close this section by taking a peek at what 'future physics' may have to say about the issues we are interested in, and about potential developments in the philosophy-science interaction more generally.

3.2 Strings and other stuff

As already mentioned, one of the greatest priorities for contemporary physicists is to construct a theory of quantum gravity able to unify quantum mechanics and general relativity. String theorists such as Greene (see his [1999]) approach this issue by regarding general relativity as a low-energy limit of a QFT where the basic entities are two-dimensional 'world sheets' existing in a background space of ten or even more dimensions. Such entities look like either closed or open strings and can split and connect. In the absence of external interactions, they obey a dynamics that leads them to oscillate. The quantum mechanical nature of the behaviour of strings implies that these oscillations take on discrete modes. Extended objects known as 'branes' are also postulated by the theory and indeed appear required to make it consistent. These can have any number of dimensions from 0 to 25. Now, string theory appears to account for certain peculiarities of QFT (for instance, particle creation is conceived of as the splitting of strings, and annihilation as strings merging) on the basis of entities that look like individual objects (although existing in a space with a number of dimensions different from the canonical three). On the other hand, that a clearly identifiable unique set of ontological posits that can place a claim to fundamentality exists in string theory can be disputed; and there are arguments to the effect that the best realist attitude towards the theory is a moderate *epistemic* realism about certain sets of relations rather than a (quasi-)classical object-oriented stance (Dawid [2007]). And, of course, it must be kept in mind that – as already explained in Chapter 1 – string theory is notoriously object of contention, and some even doubt that it qualifies as scientific. The foregoing shows that future developments in physics may provide grounds for ontological conservativeness, but they may even require a more significant degree of metaphysical revision. And that, at any rate, such developments should be evaluated with care, as their foundations, consistency and empirical significance need to be made clear before they can truly become relevant for metaphysical analysis.

Another possibility worth mentioning in this sense is one that some scientists already take seriously, and that might gain further support with future progresses in physical theorising. It is the possibility that, as a matter of fact, there is no ultimate, fundamental level of reality. That is, there is no set of things that are ontologically basic in the sense that they do not depend on anything else, and reality is instead a continuous series of ever smaller and more basic[72] entities. This possibility is clearly relevant if one is to inquire into the nature of material objects, regardless of what specific ontology one prefers. And it might become directly relevant for science, for instance if the structure of the Standard Model of elementary particles were to turn out to be just the 'surface' of a much, in fact infinitely, more complex construction. Close interaction between science and metaphysics is to be auspicated here, too. Indeed, something on this issue of fundamentality will be said later, in Chapter 5.

But enough about potential future developments in physics and in the physics–metaphysics interaction. Let us now get back to our main discussion, and look at the metaphysics of structure more closely.

4. Ontic structural realism

The idea of an ontology of structures has emerged in the context of the controversy between scientific realism and antirealism (i.e., the debate concerning the epistemic significance of scientific theories). In particular, in the last 10–15 years, much has been said in favour of a view that has come to be known as 'ontic structural realism' (henceforth, OSR). The fundamental intuition underpinning scientific realism is expressed by the so-called 'no-miracle argument' (Putnam [1975; 73]), according to which the success of science is to be explained in terms of the (approximate) truth of our theories. 'Standard' realism has it that successful theories are (approximately) true descriptions of individual objects that exist mind-independently and of their properties. The pessimistic meta-induction on the history of science (Laudan [1981]), however, threatens to sever the link between success and truth by pointing to theories that were once regarded as true, and met with some degree of success, but are now considered false. Structural realism (SR) attempts to re-establish the connection between success and truth by pointing at the structural continuity that exists between (some parts of some) subsequent theories across theory-change. SRists thus propose that we take (preserved – perhaps via some correspondence principle) structure as the (approximately) true part of such theories. Epistemic structural realism (ESR) (Worrall [1989]) intends this as an epistemological position, to the effect that we can be realists about what is described by the (preserved) mathematical structure of our theories. Ladyman [1998] introduced instead OSR, according to which not only is structure all we can be realist about, but it is also all there is. Now, what is crucial for our present purposes is that OSR is alleged to receive decisive support from a consideration of contemporary physics. This, OSRists contend, allows one to fill an existing gap between epistemology and metaphysics, by putting indications coming from physical theory and from the history of science together in a sort of virtuous consilience.

In the early years of OSR, the argument provided in support of its metaphysical component, mainly by French and Ladyman ((Ladyman [1998], French and Ladyman [2003]), was based on an alleged metaphysical underdetermination in the interpretation of non-relativistic quantum mechanics (the options being underdetermined have already been discussed in detail here). According to supporters of OSR, realists should be unhappy with metaphysical underdetermination, and consequently happy to see the problem disappear if one simply moves to an ontology of relational structure, where objects become derivative by-products of what is truly metaphysically fundamental.[73] However, we have already seen that

i) The claim that quantum particles may be completely indiscernibles might be disputed on the basis of the notion of weak discernibility;
ii) That quantum statistics can be explained without giving up on particle individuality;

and, most notably,

iii) Reasons can be given in favour of the view of quantum particles as individual objects endowed with primitive identity.[74]

In general, the claim of underdetermination can be rejected by contending that metaphysical presuppositions can and should be motivated *prior* to the evaluation of the empirical evidence, and this makes the alleged problems for object-based ontologies disappear. In addition to this, an ontology of relations only cannot in any way be regarded as a 'common core' shared by the individuality and the non-individuality views, for the latter are both based on objects and monadic properties (possessed by or constituting the former). Consequently, the doubt arises that the ontology proposed by OSRists is just an equally problematic third option.[75] For all these reasons, the argument in favour of OSR based on underdetermination in the ontological interpretation of QM can be set aside without further ado. This leads us to a second argument for OSR, more directly and positively grounded in contemporary physics.

To begin with, recall our earlier discussion of weak discernibility. Muller [2011a] emphasises that, even if those results undermine the above claim of underdetermination, it does not also undermine OSR: rather, it lends support to it. This is because objects which are only weakly discernible entirely depend on relations for their identity, and this is exactly what the structuralist claims. (Notice that the same holds if the contextualist view of identity facts is endorsed instead.) A stronger form of OSR can also be obtained by supplementing the arguments for weak discernibility (or contextuality of identity facts) with arguments to the effect that not only the identity-determining factors but *all properties* are (reducible to) relations. In this connection, OSRists customarily (see, for instance, (French [2003]) and (Ladyman [2009; Sec. 4.1]) quote or mention historical figures such as Cassirer, Born, Weyl and Eddington as authoritative exponents of the view that objects coincide with the identification of invariants with respect to the mathematics relevant to the theory. These historical remarks are then connected to more theoretical considerations. Muller [2011a], for example, argues that the (allegedly) monadic and intrinsic (essential) properties of quantum particles are, in fact, invariants of the symmetry groups that define the qualitative features of quantum mechanical systems. Others (for instance, (Kantorovich [2003] and Lyre [2004]) argue in analogous fashion that in quantum field theories, objects are secondary to structure because symmetries are fundamental in the constitution of fields (this is clearly relevant with respect to our discussion of QFT in the previous section). In general, then, OSRists interpret the mathematical nature of physical theory and the fact that all physical properties must be 'extracted' somehow from symmetries and invariants that can be traced in the formalism of the theory as clear signs that the things described by the theory and their properties are *themselves* reducible to the relations defining such symmetries and invariants.

Now, the claims made by OSRists in the present context are considerably unclear. What does it mean, exactly, that objects and properties are/reduce to invariants, or that symmetries are ontologically prior to objects? A confusion seems to lurk around between the *formal* definition of *general, abstract* properties and the *concrete* property-instances that exist in the *material* world. When one focuses on invariants and the likes, one moves at a high level of abstractness, involving the 'general properties of the general properties' of types of things. Object- and property-tokens can certainly not be 'found' there. Thinking otherwise would be like, say, expecting the actual causal features shared by coloured material things to be reducible to the general features shared by abstract concepts and words such as 'Yellow', 'Blue' and the likes. Note that the difference involved here is a difference of category: on one side, there are the properties of the abstract, formal structures used to describe the physical world, on the other the properties of concrete things. Therefore, it will not do for the OSRist to maintain that the specific property-instances of specific actual objects can be extracted via a structural analysis of the theoretical apparatuses employed for describing individual physical systems; nor that only the general/universal is real.[76]

Some OSRists just refuse to explain what distinguishes physical from mathematical structure and claim that it suffices for OSRists to indicate the relevant formal structures (Ladyman and Ross [2007; 158]). This, however, presupposes that the burden of the proof is on those who do *not* propose a radical metaphysical revision – something that appears unconvincing even independently of the specific methodological assumptions being made in the present work.

In view of this, even when argued for directly on the basis of contemporary physics, OSR appears to be a possible realist position at best, but is far from being grounded in a well-defined and plausible ontology. To be sure, then, it does not represent an especially compelling option with respect to the ontological interpretation of quantum theory in particular.[77]

5.   Conclusions

In this chapter, it has been argued that non-relativistic quantum mechanics can and should be regarded as a theory where the classical notion of individuality (with discernibility on the basis of monadic properties only) fails, but not that of individuality in general. Specific suggestions as to how to conceive of primitive identity and individuality and explain quantum statistics have been offered and defended as the most plausible ontological perspective on the theory, all the criteria for theory selection having been considered. On the other hand, it has been contended that relativistic and non-relativistic quantum field theories, while perhaps allowing for something like an ontology of individual objects, force one to undertake radical conceptual changes without certainty of success. This leads one to be more careful, and consider metaphysical revision more likely there, if not to prefer metaphysical scepticism – or at least agnosticism – in view of the shortcomings of all the extant ontological alternatives: from those less revisionary (i.e., those based on tropes) to those more revisionary (i.e., those based on 'factors' or 'processes') – the latter including the currently fashionable ontic structuralist position. Overall, the results of this first case study appear to show that the parallel study of metaphysics and physics along the lines suggested earlier is fruitful and worth carrying out, and may lead to definite positive results in some cases, and more negative conclusions in others, but always provides one with ways of identifying and evaluating options and making choices.

Having said this with respect to the nature of matter, in the next chapter we will move on to a consideration of the nature of the space(-time) in which such matter exists (or is supposed to exist).

Notes

1. Barcan Marcus [1993; 25] suggests that all those entities about which it is appropriate to assert the identity relation qualify as 'things'. Hence, it might be said that individual objects are things in Marcus' sense, and there are also non-individual objects which lack self-identity and/or numerical distinctness

from other things, and still qualify as particulars. This differentiation will play a role later.

2. If self-identity and numerical distinctness are independent of one another, one might at this point remark, then there should exist four ontological categories, corresponding to the four possible combinations, and not just two. That is, so to put it, three kinds of non-individuals. Lowe [2001] suggests that this is, in fact, the case. Equating numerical distinctness with (ordinal) countability, he distinguishes *quasi-objects* such as electrons (only countable); *quasi-individuals* such as parts (or 'portions', or 'quantities') of homogeneous stuff – think about a jug of water – (merely self-identical); and *non-objects*, such as particular qualities (lacking both self-identity and countability). While I consider it correct to claim that "analytic metaphysics can serve a valuable purpose by articulating a categorial framework which renders their possibility [i.e., the possibility of these four types of entities] perspicuous" [Ib.; 58], I do not think the examples provided by Lowe are as uncontroversial as Lowe takes them to be. That quantum particles are quasi-objects is definitely debatable, but this will be discussed extensively later in this chapter. As for quasi-individuals, Lowe seems to conflate the epistemic arbitrariness of the identification of parts of homogeneous stuff (how many molecules count as a 'portion' or 'quantity' is not uniquely fixed) with the lack of numerical distinctness in the ontological sense. (It is not the case that there is no objective number of molecules in any arbitrary amount of homogeneous stuff.) As for particular qualities – although this might essentially be a matter of definition, or at least subjective preferences with respect to the ontology of particularised properties – there seems to be no reason for thinking that particular property-instances/tropes are not self-identical and countable.

3. See, for instance, Leibniz [1704(1981)]. What 'qualitative' (as opposed to 'quantitative') exactly means is in fact less obvious than it may seem. The most uncontroversial definition is that according to which qualitative properties are those that do not involve an *ineliminable* reference to identity facts (i.e., to the primitive identities of individuals, or to properties/relations having to do with self-identity and/or numerical distinctness).

4. The converse of PII (known as Leibniz's Law) states that identity entails sameness of properties, so that if two things are identical, they have all their properties in common. This principle, unlike PII, is a logical truth no matter what properties are considered and is, therefore, philosophically uncontroversial. (Notice that seeming counterexamples, e.g., that the statue and the clay of which it is made are identical and yet have different properties, actually fail: unless one is a monist – but then the difference in properties is denied – the statue and the clay are not related by identity but only by sameness of constitution.) Incidentally, talk of 'two' things being identical may appear inconsistent: to dispel the impression, it is useful to think of PII and Leibniz's Law as, strictly speaking, being about *descriptions* of entities. PII, thus, is the principle according to which, if the descriptions of two things are absolutely equivalent, they are in fact used to refer to the same thing. Talk of 'ideal' descriptions may then be used to dispel the worry that a metaphysical principle is made dependent on epistemic limitations. This will become relevant shortly, when descriptions provided by physical theory will be systematically regarded as objective descriptions of reality.

5. Of course, this is more controversial than usually thought, for the obvious question arises concerning the individuation of spatial points. This issue will nevertheless be bracketed in what follows.

6. From now on, only 'standard' quantum mechanics – or, alternatively, a 'minimal' nucleus shared by all interpretations of the theory – will be considered. A discussion of the metaphysical consequences of specific interpretations (say, the modal interpretation) and/or alternative theories (e.g., Bohmian mechanics) would definitely be of interest, but falls outside of the scope and the possibilities of the present work.

7. An operator A on a state space is Hermitian iff:
   1. A is *linear*, that is, for all vectors u and v and any number c,
      a. $A(u+v)=Au+Av$
      b. $A(cv)=c(Av)$
   2. $<u|Av>=<Au|v>$ (see the definition of inner product below).

8. The eigenvectors of an operator A are vectors v_1, v_2, ... such that for each i $Av_i=a_iv_i$. The a_is are the eigenvalues of the operator A. If A is Hermitian, these are real numbers.

9. For simplicity's sake, we will refer to observables and to operators describing observables interchangeably from now on.

10. Defined, for any two vectors a and b (with the same origin), as $<a|b>$=length of a times length of b times cos(angle between a and b). This notation (the so-called 'bra-ket' notation, of the form $<\varphi|\phi>$, introduced by Dirac) is needed to express quantum states as vectors in Hilbert space.

11. That is, a one-dimensional subspace of the total Hilbert space.

12. Similarly, Prob $(\Delta)_O^{|\Psi>} = <\Psi|P^O\Delta|\Psi>$ gives the probability that a measurement of the observable O on a system in state Ψ yields a result in the interval Δ. In this case, the projector operator projects onto a subspace, not (necessarily) a one-dimensional ray.

13. Of course, this use of the word 'identical' is different from the philosophical one, as it *does not* imply numerical sameness.

14. Evidently, that it is the probabilities relative to particle 1 that are conditional on those of particle 2 is absolutely arbitrary, and the description can be reversed. The specific choice of values is also irrelevant.

15. That is, both for the particles constituting matter and those acting as force carriers.

16. For exchanging two particles with each other twice leads back to the original situation, so effectively giving the same result as an application of the identity operator; but it is also obvious that an operator times its inverse is equal to the identity operator.

17. Roughly speaking, the adjoint of an operator stands to the operator as the complex conjugate of a complex number stands to the complex number.

18. According to which Prob(A|B)=Prob(A&B)/Prob(B).

19. The situation is slightly more complicated for paraparticles (undetected particles that become theoretically possible if one drops the symmetrisation postulate, and that obey peculiar statistics), but this need not worry us here. It must be mentioned that French and Redhead's results have been made more general by Butterfield [1993] and Huggett [2003].

20. The possibility has been contemplated of identifying a quantum particle by referring to its *history*. Cortes [1976] argued that even knowing the entire history of a particle would not be sufficient for individuating it when it is part of a system of indistinguishable entities, because we would still be unable to 'pick out' *one specific* particle. Barnette [1978] objected that this is a merely epistemic fact, and so the possibility that histories individuate remains open. Van Fraassen [1991], though, pointed out that histories are 'empirically superfluous' in the sense that they do not add anything to the physical description of the systems in question and are in fact excluded from the range of genuine properties of the theory. More generally, it seems odd to think of histories as full-blown properties of objects, possessed by these at specific times and consequently to be included in the range of universal quantifier of PII. Rather, histories appear to be mere constructions, entirely derivative on the properties that specific objects possess at specific, successive instants.

21. Where each quantum number specifies the value of a quantity that is conserved by the particle in the dynamics of the quantum system it belongs to, and the set of all the quantum numbers of a particle exhaustively specifies its properties. For a single electron in an atom, for instance, one has a principal, an azimuthal (also called 'angular', or 'orbital'), a magnetic and a spin quantum number. Taken together, these numbers fully specify the qualities of that electron.

22. This is the working presupposition in French and Redhead's reconstruction of the violation of PII in QM. Massimi [2001] maintains that indistinguishable fermions in entangled systems cannot be attributed monadic properties, and suggests taking their properties as relational. Indeed, unlike classical entities, for which a maximally specific state description is always available, quantum entities lack such a description when entangled. And if one identifies state-dependent properties with maximally specific state descriptions, this means that there are no state-dependent properties to be 'plugged into' PII for at least some quantum systems. French and Redhead assume that the state-dependent properties of entangled quantum particles are those described by their mixed state, a move which they regard as justified by the fact that pure states and mixed states cannot be distinguished by means of observations made on one of the particles alone. But this, too, might be questioned. Since the more recent developments of the discussion, as we will see, are independent of this issue, we will, in any case, not say anything more about this.

23. The term 'entanglement' denotes, very roughly, the fact that the quantum states of two or more systems do not convey all the available information. A complete description of an entangled system must necessarily describe the entangled sub-systems with reference to each other, because there are irreducible correlations between their properties. The essential point about fermions is that EP determines that identical fermions in the same system *only* exist in entangled states.

24. To be precise, Saunders' framework is significantly different from the one employed here. Saunders mainly refers to the work of Quine ([1960] and, especially, [1976]) as it connects to that of Hilbert and Bernays [1934], and consequently considers conjunctions of formulas rather than universally

quantified expressions. In particular, in this framework it can be shown that in a language without identity and a finite vocabulary, it is a *theorem* that identity can be reduced to conjunctions of non-identity-involving formulas. Crucially, the latter must include formulas satisfied by pairs of objects but not by an object occupying both places.

25. Of course, a number of things should be discussed: for instance, whether distance relations presuppose monadic positional properties; and whether non-zero distance is necessarily irreflexive or is instead sensitive to the actual geometry of space (or space–time). But we will not get into these details, as Black's universe just serves to illustrate some points relevant for our main discussion, which exclusively concerns the ontology of quantum mechanics. For more specific discussions, see Hacking [1975] and French [1995].

26. I.e., an entangled state with a correlation among spin values and total spin 0.

27. A pure state is a state which is represented by a vector in Hilbert space; if it is also an eigenvector for observable O with eigenvalue o, then the system in that state will be measured as having value o for O with probability 1.

28. These only define probabilities – smaller than 1 – for a number of possible outcomes. It must be stressed that the theory (*via* the *Axiom of reduction*) *always* allows one uniquely to identify separate mixed states for the component particles: these are the particles' 'reduced' states, obtained by 'tracing out' one system so as to obtain information exclusively about the other.

29. Since any projection operator is Hermitian and idempotent, and so (given a projection operator P onto the one-dimensional subspace spanned by the eigenvector v) $<v|P|v>=<v|PP|v>$ (idempotence)
$=<Pv|Pv>$ (Hermiticity)
$=<cv|cv>$ (effect of the projection operator)
$=c^*<v|cv>$ (properties of the inner product)
$=c^*c<v|v>$ (properties of the inner product)$=c^*c$ (normalisation)
$=|c|^2$ (properties of complex numbers).

30. One immediate worry is that it seems odd to think of a definite relation holding between two objects without being analysable in terms of facts about such objects taken separately. However, this is exactly what acceptance of weak discernibility entails, and certainly represents a consistent metaphysical possibility – it was already entertained as such by Lewis [1986], who envisaged a scenario in which the opposite chargedness relation holds between two bodies lacking any specific charge. See Cleland [1984] and French [1989] for a related discussion of notions of supervenience and, in particular, of weak and strong non-supervenience with respect to monadic properties for physical relations.

31. It is perhaps worth pointing out that MSS may be regarded as only dealing with the *epistemological* issue of whether the quantum descriptions of physical systems supports the Quinean project of eliminating the identity sign as a linguistic primitive, not with the *metaphysical* analysis of identity and individuality. For present purposes, however, their results will be considered from a strictly metaphysical viewpoint.

32. For other classic papers, see Allaire ([1963] and [1965]), Chappell [1964] and Meiland [1966].

33. MSS may even be accused of a form of naïve realism about operators, according to which all the predicates corresponding to meaningful expressions in the formalism of the theory are automatically taken to be physically genuine. Caulton [2013] suggests that indeed the properties employed by MSS are unphysical because permutation non-symmetric. He proposes an alternative (in terms of anticorrelations of position/momentum and statistical variance). For another recent amendment of MSS' arguments based on allegedly more physical properties, see Huggett and Norton [forthcoming].

34. Dieks and Versteegh and, more explicitly, Ladyman and Bigaj propose an amendment to MSS-like reformulations of PII based on the introduction of physical systems acting as 'witnesses', i.e., as points of reference with respect to which the relevant symmetries can be broken. This approach has been put into question by Muller and Linnebo [forthcoming], who argue that 'witness discernibility' collapses into absolute discernibility. The extent to which a form of discernibility based on symmetries and witnesses is viable certainly deserves further discussion. For general discussions of identity and indiscernibility in logic and metaphysics, including some interesting results, see Caulton and Butterfield [2012] and Ladyman, Linnebo and Pettigrew [2012].

35. See the discussion in French and Krause [2006; 170–172].

36. Dieks and Lubberdink [2011] interestingly suggest that the indices appearing in the Hilbert space formalism of quantum mechanics *do not* denote individual particles. if they did, they argue, cases of superposition of positions would be possible that would be preserved when moving to the classical domain. Dieks and Lubberdink conclude that it is preferable to link the concept of particle directly to that of a localised wavepacket, as this explains the emergence of classicality whereas the traditional approach does not. This argument is relevant here, of course, as it would seem to undermine the suggested treatment of countability facts. Hence, more certainly needs to be said on these matters. For the time being, however, let us just point out that (a) it is a fact that indices appear in quantum mechanics but not in other theories, say quantum field theory; (b) Dieks and Lubberdink themselves acknowledge that the whole theory starts from the uncontroversial case of one particle described in a single labelled Hilbert space; and (c) the emergence of classical behaviour from the quantum domain is a larger issue, admitting of various solutions, and Dieks and Lubberdink's is definitely not the majority view on it and on the significance of labels.

37. In the equation g_j the number of microstates with energy ε_j (the energy of state j), k is the Boltzmann constant (relating temperature to energy), T is temperature and μ is the chemical potential (roughly speaking, a measure of the particles' tendency to diffuse).

38. This tri-partition of types of statistics on the basis of types of particles may be disputed, for example by claiming that classical systems obeying Bose-Einstein statistics are theoretically – and, perhaps, also practically – possible (see Gottesman [2007]). However, it looks as though a general distinction can, in fact, be drawn meaningfully on the basis of what is the case in nature under normal circumstances.

39. Whether and, if so, with what ontological import indistinguishability can be traced in CM is an open question that it is not necessary to delve into here.

40. The most fully worked out examples are the formalisms based on the notion of a 'quasiset', introduced, for instance, in Krause [1992] and in Da Costa and Krause [1997]. The basic idea is to posit as basic *Urelemente* so called *m*-atoms that are completely indiscernible and can be counted only cardinally. For such elements, French and Krause explain, 'identity, as it is usually understood, lacks sense; in other words, these entities are linked only by a weaker relation (≡) [indistinguishability], which mirrors an equivalence relation, but the language does not allow us to talk about either the identity or the diversity of the m-atoms' [1995; 23]. Similar ideas can be found in the 'quaset' view of Dalla Chiara and Toraldo di Francia. These authors claim (for instance, in their [1993]) that quantum particles cannot be uniquely labelled and this compels us to regard them as 'intensional-like entities', where the intensions – much in the spirit of Quine's conception of identity – are represented by conjunctions of intrinsic properties.

41. Incidentally, this would become necessary if a non-individuality-based metaphysics were also regarded as a solution to the failure of (traditional versions of) PII. For, PII seems to be applicable to any entity with properties, not just to those objects that also qualify as individuals. And, of course, it would be circular to apply PII to all property-exemplifying objects, but then add that the entities that violate it are non-individuals to which we should not have expected the principle to apply in the first place!

42. It might legitimately be argued that these questions, as well as Gibbs' paradox, point in a direction which is the opposite than Saunders suggests. Namely, to the fact that certain physical systems require a sharp switch to a different theoretical setting, i.e., the non-classical framework of quantum mechanics, where one finds indistinguishability.

43. As argued by Redhead and Teller [1992], there is also the additional difficulty that non-symmetric states in quantum mechanics appear to be in principle useless surplus structure. That is, possibilities that are allowed at the formal level but systematically ruled out at the physical level. Given this, Redhead and Teller suggest, it is much better to admit that particles are not individual objects (hence, as explained, that they cannot be in non-symmetric states), and that surplus structure only arises from the presence in the Hilbert space formalism of particle labels that have no metaphysical import whatsoever. On the other hand, it could be countered, following French and Krause [2006; 193–197], that there is a tension between the undeniable heuristic role of surplus structure in physics and the use of it as a basis for setting negative constraints on one's ontological beliefs.

44. And nothing else is added – for instance, with a view to distinguishing bundles and bundle-instances and regarding the latter as numerically unique, see Rodriguez-Pereyra [2004].

45. Things are not so simple. For a discussion of the issue represented in the bundle-theoretic setting by partial identity, see Morganti [2011] and references therein.

46. Other *a priori* arguments for considering PII a necessary truth, that we only need to mention here as they seem hardly relevant in a naturalistic

context, are the following two. On the one hand, according to Leibniz, it is metaphysically impossible for indiscernible distinct individuals to exist. (As we have seen, he intended this in the strongest possible sense, i.e., as the impossibility of sameness of monadic properties.) He believed that each individual thing is the actualisation of a complete concept in the mind of God; and, additionally, that God created the universe aiming to achieve, so to put it, the maximum variety with the least effort. And both these assumptions seem to support the idea that every existing thing must be qualitatively unique. Regardless of whether or not Leibniz's argument are sound (for example, does not the existence of two indiscernible things rather than just one object increase variety, e.g., in possible combinations of distinct objects?), however, these peculiarly theology-based arguments for PII hardly look compelling nowadays. On the other hand, consider the claim (Della Rocca [2005]) that if one does not assume PII, one has to accept implausible scenarios with many identical and co-located objects with all their material parts in common (and, notice, no properties of the whole resulting from the composition of additive properties of the parts). Were this actually the case, the consequences of not embracing reductionism would, indeed, be unwelcome: how can I know that I do not have 2000 computers in front of me and not just one? Della Rocca's argument, however, can be neutralised by pointing out that the anti-reductionist about individuality just needs to invoke as a fundamental metaphysical principle – instead of PII – the Lockean principle according to which no two numerically distinct things (of the same kind – statue/piece of bronze distinctions can be set aside as uninteresting here) can share all their parts (see Jeshion [2006]).

47. As such, it is definitely *not* analogous to the scenarios suggested by Della Rocca – see previous footnote. For similar considerations, see Hawley [2009].

48. As a matter of fact, self-proclaimed naturalists themselves have openly acknowledged the existence of valid counterexamples to PII for objects coming from science. Ladyman [2007], for instance, considers two-node graphs with no edges (mathematical systems for which an exchange of the component entities does not give rise to a new system) and concludes that these mathematical systems contain absolutely indiscernible entities. (See De Clercq [2012] for a different interpretation of the relevant graph-theoretic examples.)

49. The objection remains that reductionism is in fact the only view that does not postulate anything primitive, but just qualities all along (basically individuating universals via PII, i.e., on the basis of their 'empirical content'). A longer discussion would perhaps be in order. Here, however, we will just limit ourselves to two quick comments: (i) That qualitative uniqueness=numerical uniqueness is in any case a non-further-analysable assumption itself; (ii) The numerical identity of universals across their instances is definitely a non-negligible, and basic, assumption.

50. In this, Ladyman follows in important respects previous work by Stachel (see for instance Stachel [2002]).

51. However, see Muller [2011] for the claim that space–time points are weakly discernible.

52. For discussions, see Norton [2011], Butterfield [1988] and [1989], Brighouse [1994], Hoefer [1996] and Rynasiewicz [1994].
53. This option has been explicitly put to use to account for general relativity: see Butterfield [1989] and Brighouse [1997].
54. Based, for instance, on paragraph 30 of his *Discourse of Metaphysics*.
55. For interesting discussions of Humean Supervenience and quantum mechanics, see Darby [2009] and [2012]. More on this in Chapter 5.
56. The proposal entails, quite importantly, that in many-particle systems of identical particles, only total-system symmetric operators correspond to genuine observables (not single-particle operators, not even when an eigenvalue for the corresponding observable is possessed with probability 1). This entails, if anything, a slight modification to EEL, but the latter is certainly not an indispensable part of the theory, and just represents a possible interpretative rule, hence there is no particular reason to be worried here for the primitivist. Besides, one could insist that, in the case of entangled systems, EEL only applies to total system operators anyway.
57. The label 'Scholastic', often used generically nowadays in a pejorative sense, thus hides important differentiations. In this case, an important difference is that while Duns Scotus held that individual objects are literally constituted, among other things, by *haecceitates*, Ockham endorsed a more radical nominalism according to which the identity of things is given once those things are, i.e., identity facts are entirely supervenient on existence facts.
58. For, it can perfectly well be the case that the identity of objects is the identity of the places these objects occupy in a structure, but if nothing is said specifically about the structures themselves, the impossibility of some of them (with monadic properties exemplified as particular places) is left unexplained.
59. This should also make clear why Arenhart's [forthcoming] suggestion that – since physical theory does not provide conclusive, direct support to any view on the identity of quantum particles, and all parties to the debate agree that there must be some degree of weakening with respect to classical individuality – one may well take this to the extreme and also give up identity in the way suggested by the Received View. The problem with this is that, while *all* other approaches (try to) preserve facts of self-identity and numerical distinctness that appear fundamental, the Received View gives up exactly on this. Of course, it could be argued that the Received View is in harmony with what the theory says, and this is why it was endorsed by the founding fathers of QM. But things are more complicated than this, and perhaps a more conservative approach is in order. For a more systematic evaluation and a negative conclusion, see the next section in the main text.
60. As already made clear, there is neither the space nor the pressing need here to discuss the topic of identity and individuality in the quantum domain against the background of the whole host of alternative interpretations and/or theories. Suffice it to say in this concluding part of the chapter that Bohmian mechanics appears to allow for an essentially classical notion of individuality, as it attributes unique locations and trajectories to particles; that the ensemble interpretation (according to which the wavefunction is an abstract mathematical object that is not directly connected to real individual systems and only gives us information about the latter indirectly, by

describing ideal ensembles of systems with the same features – Ballentine [1970]) does not allow one to say anything about metaphysical of identity and individuality; and, lastly, that many-worlds interpretations, spontaneous collapse views and other interpretations that do not abandon the idea of collapse of the wavefunction do not seem to differ much from the standard view with respect to the topic at hand. More space for manoeuvre might instead be allowed by modal interpretations, in which a differentiation is drawn between the properties belonging to the *value* state of a system and those of its *dynamical* state.

61. One may feel that he or she does not share that intuition, in which case an explicit argument would be needed. One quick (and, admittedly, far from conclusive) possibility – already hinted at in the course of the discussion of the ontological nature of the identity- and difference-making relations postulated by contextualists – might be to reason as follows: first, the ontological status of an object, which ontological category it belongs to, cannot change because facts about its extrinsic relations with other entities change; in particular, an object should count as a (non-)individual independently of whether or not it is the only object that exists; but all this is only true if primitivism is correct.

62. It is important, however, to note that Newton himself explored the possibility that some more or less continuous medium is the carrier of interaction. Stein [1970], for example, interprets Newton's notions of absolute, accelerative and motive quantity as measures of centripetal force in field-theoretic terms. The centripetal force, says Stein, is for Newton "unmistakably [... analysed in terms] of field intensity: a function defined on a region of space, whose value at a point measures the tendency, or 'disposition', for bodies at that point to be acted upon". Indeed, "the field intensity is the acceleration [... and, in general, Newton's...] analysis of the situation makes an essential use of the notion of a field" [Ib; 266–267].

63. Essentially because quantum electrodynamics (the theory of the interactions between electrically charged particles and electromagnetic fields in terms of quantum fields) failed in significant ways to explain nuclear interactions and turned out to be at least partly non-renormalisable.

64. Of course, it is also an important fact that quantum field theory grounds the so-called 'Standard Model' of elementary particles.

65. In what follows, I will simply bracket the important debate about whether the philosophical analysis of quantum field theory should focus on the 'conventional' theory or rather on its algebraic formalisation. Algebraic QFT is an axiomatic framework for QFT developed in the late 1950s by Haag and subsequently further elaborated upon by him together with Araki and Kastler, and also by Wightman. It considers as basic certain peculiarly constructed sets of operators and can be 'concrete' or 'abstract', depending on whether or not specific operators on Hilbert space are explicitly referred to. See Fraser [2011] and Wallace [2011] for opinions in favour of axiomatic QFT and conventional QFT, respectively.

66. A particularly relevant physical state of systems of electrons that, under specific conditions, constitute bound pairs with no current flow. Such states are at the basis of the first microscopic theory of superconductivity, proposed by Bardeen, Cooper and Schrieffer [1957].

67. Halvorson and Clifton, in particular, reinforced Malament's claims by defending them from criticisms levelled by Fleming and Butterfield [1999] and Busch [1999].
68. Kuhlmann [2010; 103–106] argues that the results in (1) and in (2) – in particular, those of Malament and Redhead – are very closely connected, and can in fact be regarded as just one result.
69. According to which quantum field states are essentially superpositions of classical field states.
70. There are others. For instance, as suggested in passing by Bain [2011], to simply replace N with some other formal construct as the expression of the number of individual objects present in the relevant field-theoretic systems. But, of course, this is in need of further specification.
71. Rather, it seems that it is Simons who attempts to impose an a priori construction on the empirical domain as it is described by the theory. For instance, what exactly is it, in QFT, that grounds Simons' exotic talk of modes – such as, to mention just one, 'bracteal vergence' – and modal dimensions (other than the vague reference to Planck that can be found in Simons' paper.
72. But see Schaffer [2010] for arguments to the effect that what is smaller is not more basic, and that, in fact, there is something ontologically fundamental, and this something is the universe as a whole. Schaffer's 'priority monism' is discussed further in Chapter 5.
73. In their [2011], Butterfield and Caulton also argue that the permutation invariance of quantum mechanics doesn't suffice in itself for endorsing structuralism because it only points to a formal feature of the theory. However, they add, once paraparticle states are taken into account, the fact that permutation invariance holds in spite of the fact that the states are not fully (anti-) symmetric seems to suggest that invariant structure is ontologically fundamental. This argument certainly deserves further attention. It is, however, not essential for my discussion in the main text.
74. Given the methodology developed in the first two chapters and the conclusions about non-relativistic quantum mechanics formulated in this chapter, it should be quite clear why the idea of underdetermination as a ground for OSR is being rejected here; and also why the discussion of PII in QM is considered to be of particular relevance. It is true, however, that leaving things partly implicit (as it was done in Morganti [2004]) might complicate things, so it might help to add a few things to that paper here. For instance, Ainsworth [2011] claims that it is wrong to contend that PII is crucial for the whole argument, and French [2010] also points out that he (and Ladyman) never presented an argument for OSR based on the application of PII. However, the contention of Morganti [2004] was that – *since* we have a commonsense intuition that objects are individuals, and quantum objects might be individuals regardless of their (in)discernibility – any claim of underdetermination must be based on (a) an argument that the PII option and the primitive identity option (the only two considered in the argument for OSR under scrutiny) are methodologically on a par; and (b) the claim that PII is not an obvious 'winner by elimination'. In this sense, an evaluation of what the theory entails with respect to PII does seem essential. On a side note, contrary to what French thinks, Morganti [2004] did not attribute to him the view that PII is necessarily true. The point was, rather, that the

supporter of the PII-horn-of-the-dilemma has to regard the *move* from what we know of the properties of things to what is the case about their objective identity and individuality (or lack thereof) as necessary. This seems unquestionable in an empiricist setting that rules out primitive identities and yet accepts metaphysical discourse as more than fiction, and accepts reductionism about individuality. French also misunderstands the use of Lowe's concept of quasi-individuality, not intended, as he suggests, to point to lack of self-identity, but rather to the possibility of regarding pairwise discernibility and permutation symmetry as a 'collective' property (in the sense of Hilborn and Yuca [2002]), of certain types of entities that count as individuals *at least* in a minimal sense; and of Loux's version of transcendental individuality, which was used as an example of how the alleged underdetermination might be broken by opting for the non-PII horn of the dilemma.

75. This argument can only be neutralised by claiming that what exists are physical structures determining object-places that may (or may not) be filled by objects, the latter being either individuals or non-individuals. Here, however, we are interested only in the general features of extant arguments in favour of OSR based on quantum theory.

76. OSR would be a form of realism about mathematical, not physical, structure – this is a possible philosophical position (see Tegmark [2007]), but not the one that supporters of OSR generally wish to defend. For a recent variant of OSR which clearly moves at a high level of abstraction, and is especially formulated with QFT in mind, see Peterson and Silberstein's [2010] 'relational blockworld' view.

77. This conclusion extends also to the 'less radical' variants of OSR that preserve objects alongside structures. These will be discussed further in Chapter 5.

4
Space and Time

1. The metaphysics of time

One traditional theme in metaphysics, at least as old as that concerning the nature of material objects, has to do with the ontological status of time and space, i.e., of those entities supposed to constitute the stage, as it were, in which material actors perform. Indeed, upon analysis, our intuitions about the basic characteristics, and even the existence, of space and time turn out to be problematic, as a range of issues emerges that deserves careful philosophical scrutiny. Some of these issues concern equally space and time, others are specifically about time. It is these issues that this chapter will be concerned with.[1]

Essentially, there are two dichotomies that the metaphysician who is interested in the nature of space and time has to deal with.

One coincides with the opposition between the view called *substantivalism* and what is known as *relationism*. According to substantivalists about space and/or time, space and/or time exist independently of physical objects and processes and are prior to them. As a matter of fact, they constitute the container in which objects exist and physical processes occur. According to relationists, instead, space and/or time depend on physical objects and events: they are derivative on, and even reducible to, relations between things. Putting the same point slightly differently, for the substantivalist an ontological catalogue of what exists as a fundamental entity includes (portions of) space and/or time, while for the relationist it does not. Supersubstantivalism, then, is the extreme substantivalist view according to which material objects can be somehow reduced to portions of space or space–time. With respect to this first dichotomy, it is usual to believe that it allows for two possibilities only, i.e., substantivalism and relationism about *space–time*. However, it is obvious that, logically speaking, the distinctness of space and time leads to four possibilities. Thus, it is necessary to see exactly what physical and/or philosophical arguments there might be for/or against the unification of space-time.

The second dichotomy to be introduced concerns time only. It has to do – roughly![2] – with the question of whether only the things that exist *presently* are real, or – contrary to appearances – past and (perhaps) future events are on the same ontological footing as present ones. The debate here is between *presentists*, who support the former view, and *eternalists*, favouring the latter (but we will see that there are nuances to be taken into account).

Let us now look in more detail at these dichotomies and the surrounding issues, starting from the controversy between substantivalism and relationism.

1.1 Substantivalism v. relationism

It is customary to trace the origin of the debate between substantivalists and relationists in the work of Newton (1642–1727) and Leibniz (1646–1716), and in the exchange between the latter and the followers of the former (mainly Samuel Clarke (1675–1729)). In the *Principia*,[3] Newton distinguished an *absolute* space, an intrinsically 'homogeneous and immovable' entity, from *relative* space, which is basically our mutable reconstruction of space based on measurement procedures with respect to arbitrary reference bodies. Something similar Newton thought to hold for absolute time as opposed to relative time. He maintained that only the former, by its own nature, passes equably without relation to anything external and thus without reference to any occurrence of change or way of measuring it.

To begin with, let us focus, as is perhaps customary, on what Newton had to say about space. In virtue of his distinction between absolute and relative space, Newton also believed in absolute places (the portions of absolute space occupied by material bodies) and in the 'true motion' of things, identified with their motion through absolute space. These views were developed by Newton primarily in opposition to Descartes (1596–1650), who considered motion relative to specific bodies and, more generally, followed the then traditional view that the universe is a material plenum.[4] Paradigmatic in this sense is, of course, Newton's notorious bucket experiment. If we picture a bucket suspended by a rope which starts rotating around the rope's axis, at first the bucket and the water are at rest with respect to each other, then the bucket rotates, but the water is at rest (bucket and water are in motion relative to each other), but ultimately the bucket and the water are mutually at rest (the water moving with the same velocity as the bucket). Since in the last phase of this process the water certainly moves – as witnessed by the fact that the water's surface becomes concave – but water and bucket are at rest in relation to one another, there is something that must be explained but cannot be explained in terms of merely relative motion. Hence, Newton's reasoning goes, one must have here something that is only definable against the background of absolute space.[5] (Nowadays, these views and arguments are normally associated to substantivalism. This is correct, but it must be pointed out that Newton – based on what he regarded as the proper definition of a substance as a self-subsistent entity capable of acting – only stated that space and time are 'closest' in nature to substances).

With respect to time, Newton argues in favour of the distinction between absolute and relative time by pointing out that astronomers use an equation of time to correct for inequalities in the solar day, i.e., in the rotation of the Earth around its axis. This he took not only to show that astronomical practice presupposed a uniform time, more fundamental than that reconstructed on the basis of actual motions of specific bodies. But also, more importantly, that the existence or non-existence of uniform motion is a contingent matter, and thus the nature of time as we conceive of it cannot depend on the motion of actual bodies.

On the other extreme, one finds relationism. A relational view of time can be traced back to Aristotle. In the *Physics*, and in particular in sections 10–14 of Book IV, Aristotle explicitly argues that time is distinct from change,[6] but existentially dependent on it, that is, on relations between distinct events. In particular, for Aristotle, time is "a number of change with respect to the before and after" (219 b 1–2). Leibniz makes a similar point when he says that "instants, consider'd without the things, are nothing at all; ... they consist only in the successive order of things" ([1704(1956), third paper, section 6, emphasis added]). Of course, Leibniz was a relationist also about space. For example, he stated: "I hold space to be something merely relative, as time is; ... I hold it to be the order of coexistences, as time is an order of successions" [Ib.; 25–26]. The motivation for this particular sort of relationism, applied to both space and time, was for Leibniz that both space and time as containers for physical objects and events are superfluous (for instance, true motion is, according to Leibniz, reducible to the active forces inherent in certain bodies); and that, so understood, they would entail too many unrealised possibilities, a choice among which could not be grounded in any way.

Of course, we are talking here about the well-known observationally indistinguishable shifted universes that Leibniz used to refute Newton and his followers. Suppose substantivalism is correct. Why, asked Leibniz, did God decide to create a universe oriented in a specific way rather than another, or at a specific position in time and space rather than any one of many others? A version of these worries which is particularly relevant nowadays consists in the so-called 'kinematic shift argument', whereby two possible worlds only differ with respect to the value of the absolute velocity (i.e., the rate of true motion) of each material object inhabiting them. This possibility leads the Newtonian to posit certain surplus physical structure – in particular, in principle unobservable absolute velocities – which can simply be dispensed with by opting for a relationist framework.

Or so it would seem, at least. Historically, Leibniz's theologico-metaphysical arguments actually lost against the more empirically grounded considerations offered by Newton. Putting it in general terms, substantivalism about space and time came to be regarded as correct, as an absolute background appeared necessary for physical theory to work. In other words, Leibniz's relationism ultimately failed to prevail on the Newtonian perspective due to the fact that Leibniz and his followers did not manage to define an adequate physics, one not requiring inertial structure and a temporal metric (see, e.g., Pooley and Brown [2002; section 4]).

This also holds true for the work of Mach, who (especially in his *Science of Mechanics* (1883)) attempted to elaborate on Leibniz's insights and vindicate the relationist stance. Mach started from the observation that, by positing an absolute space and an absolute time, Newton departed from his own key methodological tenet according to which one should not go beyond observational facts: for, clearly, one cannot observe absolute space and/or time. Mach tried instead to fully implement this Newtonian principle. This led him to dispense with absolute space and time altogether, and try to make do with relative distances and in principle observable magnitudes only. For instance, Mach interpreted Newton's bucket experiment so that what plays a crucial role is not absolute space but, rather, the relation between the water in the bucket and the fixed stars. Generalising, Machian relationism suggests a systematic replacement of all references to absolute space and time with reference to systems of actual objects.

However, Mach's project cannot be regarded as a success. Even though people such as Reichenbach went as far as to argue that Einstein's theory of relativity fully vindicated the Machian perspective, the workability of a fully relational physics remained, in fact, doubtful to say the least. Indeed, in spite of the fact that they were designed to be generally covariant (i.e., such that natural laws can be expressed by equations valid for any coordinate system) and thus 'background independent' (i.e., deprived of a fixed spatio-temporal background giving rise to inertial effects), Einstein's theories appear to imply substantivalism about *space–time* intended as a four-dimensional whole. At least since the 1960s this is, in fact, the dominant view in the philosophy of physics. True, as we have already seen in the previous chapter, if in general relativity space–time points are conceived of as individuals in the canonical sense, then the hole argument represents a problem. In particular, if one has two distributions of metric and matter fields (i.e., space–time points and material objects in them) related by a hole transformation,[7] it seems that substantivalists must maintain that the two distributions represent two distinct physical systems. But such distinctness is in principle empirically undetectable, as the two distributions are observationally identical, and the laws of the theory cannot choose between the two developments of the fields into the hole. However, as mentioned in an earlier section, it is possible for the substantivalist to 'weaken' the identity of space–time points in such a way that diffeomorphically-related models are regarded as physically identical, so as to rule out the possibility of merely haecceitistic differences between distinct space–time models. (All this presupposes the rejection of the view that the entire debate between substantivalists and relationists is no longer relevant – see Rynasiewicz [1996] and the response in Hoefer [1998]. Such view is, nevertheless, important and not isolated in the philosophical community, and thus deserves more attention than we can give it here).

This is not the end of the story, however. Indeed, a serious re-evaluation of relationism is ongoing (see Pooley and Brown [2002] and Pooley [2001]). On the one hand, the relationist ontology remains more parsimonious and, obviously enough, also steers clear of Leibnizian worries about indiscernible universes – nowadays, of course, more problematic from the methodological perspective than from a theological point of view.[8] On the other hand, it now looks like relationism can be developed further. In this context, it is useful, in particular, to look at the recent elaboration on the Machian approach to physics proposed by Julian Barbour. Such an elaboration is the most promising attempt to revive the relationist ideas underlying the work of both Leibniz and Mach, and thus to vindicate relationism both as a metaphysical doctrine and as a perspective on physical theory. Moreover, it seems obvious that Barbour's approach has a number of consequences that cannot but be of interest for the metaphysicians, especially for what concerns the nature of time.

Barbour's fundamental insight (most fully developed in his [1999]) is that the Machian idea of grounding the whole of physics in relations between (observable) quantities – in particular, relative distances between material objects – can (and should), in fact, be upheld. According to Barbour, the only things that exist are, indeed, *configurations* of physical systems, that is, of interrelated objects and properties. Understanding the physical behaviour of things, Barbour argues, requires a generalised principle of least action[9] which sets constraints on the way configurations 'follow' each other. Crucially, one configuration does not follow another in the sense that they are ordered in an actual linear sequence (i.e., temporally) but, rather, in the sense that there are 'special connections' between distinct entities *that are all given at once*, much like the individual pieces of a completed puzzle. In particular, according to Barbour, the universe is the collection of all possible configurations (that make physical sense), and these configurations satisfy a criterion of 'best matching' – basically, a generalisation of Pythagora's theorem that minimises the overall difference among pairs of distinct configurations with respect to all the quantities appearing in them.

But how does Barbour deal with the abovementioned problem that non-relational structure appears needed in order to fully account for all the relevant facts about physical systems? His strategy is to add to the notion of best matching a simple, yet significant, presupposition: namely, putting it in Newtonian terms, that the total angular momentum of the universe as a whole is zero. Indeed, if the latter is the case, then relative quantities turn out to be sufficient for a complete physical description of physical systems and their evolution, and so one can dispense with the Newtonian 'container' altogether.[10] Against this background, together with his collaborator Bruno Bertotti (see, e.g., Barbour and Bertotti [1982]) Barbour showed that, once general relativity is formulated as the dynamical theory of the geometrical features of space coupled with that of matter fields, it indeed turns out to have a purely Machian nature, i.e., not to require any further treatment in order to satisfy the relationist desiderata. More specifically, when intended in the sense of geometrodynamics on superspace (the latter being the configuration space constructed on the basis of the set of 'acceptable' geometries of space), general relativity can be regarded as the theory of the relationships between configurations, the relevant configuration space being an entirely relative one.

Next, Barbour put quantum considerations into the picture and argued that quantum theory, and consequently quantum gravity, can and should be formulated in terms of instantaneous relative configurations. This, he explained, requires one to use the time-independent Schrödinger equation[11]

$$E\Psi = \hat{H}\Psi$$

and consequently conjecture that the quantum wavefunction of the universe explores all possible configurations 'at once'. This means that quantum probabilities do not describe anything like future measurement outcomes, dispositions or future-oriented propensities, but rather determine – by being extended over the entire collection of configurations as a 'mist' of varying intensity – the 'degree of reality' of each configuration (in a sense that will become clearer shortly). Indeed, Barbour defines and endorses a 'many instants' interpretation of quantum mechanics, which turns the Everettian many-worlds line of thought into the idea that all the relevant possibilities – Barbour calls them 'Nows' – exist together and are, in fact, individual and non-interacting parts of the same, unique puzzle in configuration space.

Here's where the important element for our present purposes emerges. Following the route sketched above entails that, in the case of the universe as a whole, the Wheeler–DeWitt equation

$$\hat{H}(x)|\Psi>=0$$

must be used, which clearly seems to encode the information that the wavefunction of the universe is constant. That is, *that the universe does not change*. While regarded as mistaken or even meaningless by many physicists, then, the Wheeler–DeWitt equation turns out to be natural within, and instrumental to, Barbour's relationism. For, it ideally expresses the key idea that what seems to be a sequence of physical arrangements in time is instead a path in a *timeless, static configuration space*. It is this, then, that the abovementioned idea that all possible configurations 'exist together' as 'many nows' truly means.

There is, of course, a lot to be said about Barbour's Machian approach to physics and its philosophical import. For instance, as we have seen, the geometrodynamical formulation of general relativity turned out to be the Machian version of the theory that Barbour was looking for, and in it the metric based on best matching between configurations makes it possible to ground the dynamics exclusively in relative three-dimensional configurations. This, however, has been taken (Pooley [2001; Sec. 3.2]) to lead to indeterminism. The idea is, essentially, that there are many different sequences of configurations of the desired type (i.e., satisfying a least action principle in geometrodynamics) that can be 'extracted' from the canonical four-dimensional relativistic space; but such sequences constitute observationally-indistinguishable 'histories'; moreover, two of these sequences can be identical up to a point and radically differ afterwards, and so the specification of an initial sequence-segment is, in fact, not sufficient for predicting the rest of the evolution of that sequence. (Clearly, this problem is analogous to that arising in the context of the hole argument discussed earlier.) This suggests, says Pooley, that the traditional four-dimensional formulation of general relativity might, after all, be regarded as preferable to its Machian reformulation. However, the surplus degree of freedom that gives rise to this issue might be disposed of by formulating the theory in so-called 'conformal superspace', which allows one to identify families of sequences in such a way that, in effect, a unique curve in relative configuration space is individuated in the relevant cases, and the threat of indeterminism is consequently neutralised. This suggestion is very recent, and still under examination (see Barbour and Ó Murchadha [2010]), but so-called 'shape dynamics' seems to represent a promising avenue of research for the supporters of the Machian approach to physics.

Other problems for Barbour concern his way of extracting our experience of time from an essentially timeless reality, and how timelessness itself should be understood exactly. However, these are particularly important issues in the present context and will therefore be discussed in detail later. For now, what is important to point out is that Barbour's reconstruction of physics is based on a genuine form of relationism and a sharp separation of time (which he then eliminates) from space. From this, the possibility (if not the need) of a separate metaphysical treatment of space and time immediately follows, together with a number of interesting metaphysical consequences and ramifications – especially as far as the metaphysics of time is concerned.[12] Before discussing Barbour's relationism further and seeing where it leads, however, we need to say more about the second dichotomy this chapter focuses on, namely that between presentism and eternalism.

1.2 Presentism v. eternalism

As mentioned in the introduction, while the substantivalism/relationism dichotomy applies equally to space and to time, there is another set of issues – revolving around the ontological status of the present as opposed to that of the past and the future – that pertains exclusively to the philosophy of time. Of course, here too the possibility that space and time constitute an indivisible whole cannot be ignored, especially in view of the fact that it appears sanctioned by relativity theory. Still, at the level of pre-scientific, a priori reflection, certain questions specifically concerning time seem to be clearly meaningful. And it might plausibly be argued that a clarification of them is essential for one's subsequent understanding of the relevant physics. Let us then focus on the temporal dimension only, starting with some general definitions and then attempting to see precisely what the debate consists in.

It is customary to begin this kind of discussions with two distinctions: one between the *A-series* and the *B-series* and another between *tensed* and *tenseless* theories of time.

The notions of A-series and B-series were introduced by the British philosopher John M.E. McTaggart (1866–1925), based on the two ways in which temporal elements figure in our language. Indeed, we can either say that events are earlier than, simultaneous with or later than one another (the B-series); or that a particular event is past, present or future (the A-series).[13] Correspondingly, our language can be tensed or tenseless, depending on whether or not it involves the notions of past, present and future as denoting fundamental facts about reality. This leads directly to the second distinction. Tensed theories (or A-theories) are those philosophical conceptions of time that draw ontological distinctions between past, present and future, and regard being past, present or future as non-analysable, objective features of things and events. Among these, as mentioned already, *presentism* states that only presently existing things are real. According to *possibilism*, or the *growing block view*, the past is different from the present but is nevertheless fully actual – the present being a continuously moving edge of an ever increasing 'amount of real stuff', and only the future counting as merely possible, that is, not real.[14]

Tenseless theories (or B-theories) are instead those according to which all times are ontologically on a par. That is, time is very much like space in that there is no objective, absolute property such as being present, exactly in the same way in which there is no objective, absolute property of being here. That is to say, 'now', like 'here', is an indexical, and the distinction between past, present and future is only relative to specific events. This leads to the *eternalist block view*, which naturally regards space and time as aspects of a unique four-dimensional whole. Clearly, tenseless theories deny that time flows in any sense involving a passage from non-existence to existence, while tensed theories take as their fundamental starting point the idea that time actually flows and there is a genuine asymmetry between the present (and the past/future) and other times in terms of reality and existence.[15]

In a nutshell, the connection between the two distinctions just illustrated is the following. Tenseless theorists put their emphasis on the B-series, for they regard the fixed, objective ordering of events as the most important thing; tensed theorists consider the A-series more relevant instead, because they believe in an ontological asymmetry between things that exist (or events that take place) now[16] and other things (or events). One may infer from this that the B-theory and eternalism are essentially the same thing, that is, that there is only one way of conceiving of time if reality is thought of as a block characterised by temporal relations, and past, present and future events/objects are regarded as ontologically on a par. Correspondingly, one may be led by what we said about presentism and the A-theory to think that the idea of the present enjoying a privileged ontological status can only be formulated in a genuinely A-theoretic setting. This is not the case: one can endorse the idea that reality is an unchanging four-dimensional block while also being a presentist. This can be done by conjecturing that there is something like *objective becoming*, but it is not grounded in ontological differences between present times and other times. This is possible if presentness is an objective feature of events but gets attached, as it were, 'from the outside' to particular instants (i.e., points in the block universe) and the entities that occupy them. This is the *moving spotlight theory* of time (see Skow [2009] and, for a related view, Brogaard [2000]).[17] We will discuss the moving spotlight theory and the other alternatives later. For the moment, let us get back to our general introduction of the issue, bearing in mind the precise way in which the basic differentiations are to be drawn.

A first, fundamental problem for those interested in the presentism/ eternalism debate is that it is not in any obvious way more than a mere pseudo-debate. Recall that, as a first approximation, we defined presentism as the view according to which 'only the things that exist presently are real'. Some authors (see, for instance, Meyer [2005] and Dorato [2006]) argue that this does not work. One way of putting their objection is as follows. Assume (as we have done so far) that the entities to be considered ontologically basic when it comes to the content of space and time are physical *events*. Now, for events, to be real means to occur (i.e., it is essential for events that they exist by taking place). But events occur at particular points (or regions) in (space-)time. Hence, to say that only present events are real is tantamount to saying that only present events occur when they do, i.e., now. Thus, for future events not to be real just means that they do not occur now. But this is something with which eternalists do not disagree. On the other hand, eternalists cannot be interpreted as maintaining that everything exists 'at once' in a real 'block', given once and for all. For, this would mean that all events occur simultaneously, which is impossible: for any point in (space-)time, as the eternalist is happy to concede and the presentist cannot deny, there are events that have occurred, events that are occurring and events that will occur. Another formulation of the objection goes like this: if one claims that everything that exists now, either 'exists' means 'exists now', or it means 'existed, or exists now or will exist', for the verb 'to exist' is necessarily tensed. But in the former case, one makes a trivial claim that both presentists and eternalists agree with, while in the latter case, one makes a false claim that neither presentists nor eternalists will agree with (see Meyer [2005; 213–216]).[18] This also seems to apply to claims such as Crisp's [2004], according to whom presentists and eternalists genuinely disagree over the truth-content of claims such as 'There is something that was the Roman Empire and is no longer present'. (True, says Crisp, for eternalists but not for presentists, who simply exclude the Roman Empire and analogous things from the domain of ontological commitment). For, the initial 'there is' clause simply perpetuates the ambiguity.

Sceptics about the meaningfulness of the presentism vs. eternalism dispute also reject the idea (see Sider [2006]) according to which the presentist can legitimately distinguish between ontologically-committing and non-ontologically-committing uses of the existential quantifier.[19] Lastly, they argue that having recourse to the indefiniteness of truth values for future-tensed propositions may re-establish a difference between the two positions, but this has mere semantic, not ontological, import.

These objections, however, are not conclusive. For it looks as though presentists can, in fact, show that their position is well-defined and distinct from the eternalists', so proving that the sceptic arguments point to merely linguistic difficulties.

To begin with, the intuition that the presentism/eternalism dichotomy is substantial can be given support by comparing space and time. Is not the disagreement over whether (a) times other than *now* are analogous to places other than *here*, hence not special in any ontologically-significant sense (eternalism) or, instead, (b) time is different from space because there is an objective now, and the present is unique (presentism), a meaningful and genuine one? After all, the four-dimensional block intuition is exactly that all points of (space-)time are ontologically on a par, even though this does not entail that everything occurs at the same time. Presentists, though, can (and should) endorse exactly the opposite view and claim that everything exists (is real, occurs) at the same time, and in such a way that certain times are objectively different from others. In connection to this, the suggested analogy between space and time in a presentist framework can be further brought out by thinking about time-travel. Indeed, presentists suggest that the present temporal location is unique exactly in the same sense in which there would only be a 'here' for a point-like subject existing in a universe with only one spatial point (or, at any rate, only a limited spatial region to occupy), hence no other point (or region) to move to from there. It would thus seem that under eternalism, but not under presentism, can the possibility of time travel, i.e., of travelling from the present to times different from it, be countenanced.[20]

One way to lend further support to the meaningfulness of the distinction between presentism and eternalism is by drawing an analogy with the debate about possible worlds. When it comes to possible worlds, the Lewisian modal realist identifies merely possible things with concrete things (i.e., things existing in space and time) that exist in a world which is as real as ours but not spatio-temporally related to it. And this is a claim that is neither trivially true nor obviously false. The actualist opposes the Lewisian view by saying that all concrete things are spatio-temporally related as only one world, the actual one, exists. And this, too, appears to be a clear, non-trivial claim, and one with which the Lewisian realist will definitely disagree. Now, the presentist can similarly be understood as claiming that everything concrete is *merely spatially related* to everything else (that is, that all physical things coexist in a three-dimensional 'sheet'). By contrast, the eternalist will be interpreted as making the claim that there are objects and events that are ontologically on a par (i.e., they all exist in the only sense of the term) and yet are not merely spatially related.[21] Notice that each party is talking about everything concrete unrestrictedly, and that the claims being made are clearly distinct. On the one hand, one finds a 3D collection of elements at some spatial distance from each other, and on the other a 4D block of entities which may be more than just spatially separated. Here too, then, we seem to have the formulation of a substantial disagreement.

Let us now consider one last point, having to do with the truth value of claims about the future. While it is true that mere indefiniteness of truth *values* does not guarantee an ontological asymmetry between present and future, there is more to say about this than sceptics think. In particular, it would seem that the presentist can, in fact, successfully defend his or her position by focusing on truth-*makers* rather than truth values. Indeed, under eternalism, once what point of space–time counts as the present is fixed, the truth values of claims about the future may well be indefinite. But eternalists *cannot* say that the truth-makers for those claims, i.e., the events occurring at the relevant (future) times are not part of the 'ontological catalogue' of reality. Indefiniteness is then, necessarily, a merely epistemic matter in an eternalist setting. In a presentist context, by contrast, claims about the future may well have a definite truth value (say, because they follow deductively from claims that are true now – think about a Laplacean omniscient being in a world governed by deterministic laws). But it is still the case that the truth-makers for those claims do not exist, and thus it is always possible that those true claims about the future never turn into true claims about the present. Not surprisingly, detractors of the substantiality of the debate are forced to insert parenthetical remarks like 'unless the whole universe comes to an end' when critically analysing presentism with respect to truth and truth-making (see, e.g., Dorato [2006]). Indeed, only in a presentist context, *not* in an eternalist context, is the possibility that all claims about the future remain without a determinate truth value (or lose their previously determinate truth value) a genuine ontological possibility. Those parenthetical remarks are, therefore, far from subsidiary and, in fact, may be exploited to express an essential point of divergence.[22] (It is perhaps worth pointing out that the truth or falsity of eternalism and the truth or falsity of determinism are completely independent of one another; it is possible both that presentism is true and the fundamental laws of nature are indeterministic and that we live in a genuine four-dimensional block, paths across which are nevertheless governed by indeterministic laws).[23]

One may or may not think that the recently proposed view (Tallant [forthcoming a]) that presentism should be understood as the view according to which being present *is the same as* existing, i.e., presentists should *identify* being present and existing, represents a useful addition to the foregoing considerations. On the positive side, this explicit identity claim does seem to be effective in expressing the fundamental presentist insight, which is squarely about how many things populate the ontological category of concrete things, i.e., exist. On the negative side, it is unclear whether such proposal truly sidesteps the linguistic problems considered earlier: does not 'presence is existence' translate into 'Whatever is present exists', with the ensuing ambiguities between different renderings of the tensed verbs?[24] We will set 'existence presentism' aside for the time being.

Having argued in favour of the substantiality of the debate between presentists and eternalists, let us now move on to another potential source of worry. It consists in the fact that, even granting that it is genuinely distinct from eternalism, presentism might be unable to do all the work that a theory of time is expected to do – so turning out not be a serious option anyway. This contention has been formulated on the basis of various considerations. The most serious among these seem to be the following two:

i) The truth or falsity of claims about the past: if every truth has a truth-maker, it presupposes the existence of something; how can

then claims about the past be true (or false) if only presently existing things exist (this is the so-called 'grounding objection')?
ii) Cross-temporal relations (i.e., relations between entities existing at different times): if only presently existing things exist, how are relations with non-simultaneously existing relata (e.g., comparative relations such as '*x* is heavier than *y* was exactly two years ago')[25] possible?

Starting from truth and the grounding problem, presentists have dealt with the latter in a number of ways. Some authors postulate as truth-makers for true claims about the past states of affairs involving present past-directed properties of the form 'being such that it was the case that...' (see Bigelow [1996] and the more extended discussion in Crisp [2007]). Another option is to postulate the existence of temporal distributional properties: Cameron [2011], for instance, claims that properties like being P, then Q, then R... just need to be coupled with specific times to do the trick for presentists.[26] Other philosophers argue that all the problematic truth-makers are presently existing abstracta, and since presentism is a claim about concrete, spatio-temporally-located entities only, the grounding problem is thereby immediately solved. (See Bourne [2006; chapter 2], for instance, where it is argued that all times are sets of propositions, but only one of them has a concrete realisation.)[27] Lastly, presentists can legitimately reject a truthmaking theory that requires every truth to have a presently existing truth-maker: Baia [2012], for instance, plausibly contends that presentists can endorse a view of grounding such that 'every true proposition depends for its truth on how the world was, is, or will be'[28] (see also Tallant [2009]).

Responses to (ii) generally go along similarly deflationary lines. Whenever a relation involves non-present relata, the presentist can analyse it in terms involving (present concrete entities and) abstract entities and, in case, atemporal relations among the latter (see De Clercq [2006], where ideas first hinted at by Prior in the 1960s are developed in detail). Whether one interprets the resulting relations as mere conceptual constructs or as committing to the mind-independent existence of abstracta is open to discussion, but the basic idea should be clear enough.

Enough for what concerns presentism. As for eternalism, its much richer ontology seems to make it immune to objections like those considered so far. The problem with it is, essentially, that it seems in conflict with entrenched commonsense beliefs. In particular, as a B-theory eternalism entails that the future is not open (i.e., that it is as fully determinate as the present), and this seems to entail in turn that we are not genuinely free agents. For, at most, we do not *know* what will happen, but what we do now affects what will happen only as a link in a causal chain that, being real in all of its parts, can only evolve in one way – such evolution not involving any strong sense of becoming. To those who draw from this the conclusion that we are not morally responsible, it might be responded that free will does not require the openness of the future, as our actions remain genuinely ours, that is, actions that – albeit in the context of a more fundamental series of events that is necessarily (in the nomological sense of necessity) what it is – are caused by us and make us responsible for particular events in the world. It goes without saying that this form of compatibilism with respect to the future (as closed) and free will is open to objections as it doesn't seem to fully vindicate the initial intuitions. It is also a fact, however, that, if anything, considerations such as those just made concern not the internal consistency of eternalism, but just its *prima facie* appeal.[29] Without entering into the intricate discussion of the issues surrounding human action and the metaphysics of time, therefore, we can postpone any further comparative assessment and conclude our discussion by suggesting that the presentism/eternalism dichotomy – much like the substantivalism/relationism dichotomy – is a genuine one that, as such, deserves careful philosophical scrutiny. Bearing the foregoing in mind, let us now see, as in the case of substantivalism and relationism, how presentism and eternalism fare with respect to the input coming from science, and in particular contemporary physics.

Independently of whether one is a substantivalist or a relationist, classical mechanics presupposes an absolute present and an objective distinction between it, the past and the future. More particularly, in Newton's world, there exists a way of describing how things are and how they evolve in time that would be accepted by all observers, that is, a privileged frame of reference, whose temporal dimension is *the* temporal dimension of the evolution of the universe – absolute, unique and universally valid. Of course, though, nowadays it is Einstein's theory of relativity that we have to look at in order to describe the spatial and temporal structure of reality. And in addition to merging space and time into a unique four-dimensional entity, relativity also puts into doubt exactly the notions of an objective, absolute time and of an objective, absolute temporal becoming. This is because in relativity:

a) The speed of light is a limit, nothing can travel faster than light;
b) Such speed is invariant across frames of reference, that is, is always observed to have the same value regardless of the speed and direction of the observer (and of the direction of light itself);

c) Simultaneity can only be defined relative to a frame of reference based on the speed of light;
d) Therefore, observers in different frames of reference (i.e., in different states of motion) can disagree on whether two events are simultaneous, one earlier than the other or vice versa.[30]

Since it would seem absurd to make the existence of the things one observes dependent on the state of motion of the observer, and it would similarly appear implausible to think that two events that appear simultaneous to an observer nevertheless fail to coexist in the perspective of that observer, one is led to the conclusion that past, present and future are ontologically on a par (i.e., the B-theory is correct, and the A-theory is not). For, then, the disagreement we just highlighted would become not one about what is real or exists but rather a mere difference of perspective on the same, objective reality (ontologically uniform, as it were, at all its points). Notice that, crucially, the theory intimates that all frames of reference, hence all judgments about simultaneity, have the same dignity, and thus there is no reason for insisting that, the evidence notwithstanding, there is something like an absolute, objective present at a supposedly more fundamental level. In a word, the difficulty raised for presentism by the theory of relativity appears to be ontological, not only epistemic. What can the presentist say at this point?[31]

A first option is to accept the four-dimensional block view of space–time while insisting on the objectivity of the present. This, of course, is what the abovementioned moving spotlight theory of the present does. Skow [2009] argues that the notion of a 'spotlight present' indeed allows to make presentism consistent with relativity. In a nutshell, Skow first considers a classical setting and suggests that temporal operators (which, notice, are in actual fact primitives of the theory) be analysed in terms of quantification over points in a so-called 'super-time', the dimension along which the moving Now moves as it illuminates different events – intervals in supertime directly corresponding to temporal units separating the 'being now' of a certain event from that of another. Then he constructs a 'Minkowski superspace-time', whereby intervals are constructed so as to mirror space–time intervals in physical reality. So doing, Skow believes, a satisfactory analysis of the process thanks to which different events/world-slices become present is provided.

This is, no doubt, an interesting proposal. Notice, however, three things. First, the moving spotlight theory assumes a four-dimensional view, hence sacrifices the commonsense idea of an open future anyway.[32] Secondly, even though from the perspective of each point in space–time there is only one point that qualifies as the present, on the proposed construal it is, in any event, still the case that different points are to be regarded as present from different perspectives: there is no unique, *shared* present. Third, a key intuition underlying presentism seems to be that once we are given everything that (unrestrictedly) exists, we are *ipso facto* given something (a three-dimensional 'slice') that we can identify with the privileged moment that we call the present. Nothing like this happens in the moving spotlight theory, where instead presentness must be added – somewhat artificially – to the whole of reality intended as a structured collection of objects/events. This, it might even be contended, is exactly what a serious approach to relativity invites not to do.

The more modest, but rather natural, idea that has been explored in the past is that of redefining the present in a way that, while preserving our basic commonsense intuitions, is compatible with relativity. This essentially amounts to dropping the assumption that the present should be identified with an extended, all-encompassing three-dimensional slice, while sticking to what seems minimally required for making sense at least of our *experience* of time. This is not easily achieved. One fairly 'extreme' alternative in this sense is to identify the present with a *point*, the one occupied by the observer. This guarantees invariance, as the contents of any point in space–time are the same from the perspective of any other point. However, it does not account for our actual experience and knowledge of reality, which seems to tell us that we share the present at least with what exists in our immediate proximity. Insisting on invariance, one might identify the present with the *surface of the past light-cone* of a given observer (i.e., the set of all events that are connected to the observer by a ray of light),[33] or with the *region containing all events that are not causally connectible* with respect to the observer.[34] In these cases, too, however, the price to pay seems too high: for, in the former case, our notion of the present should be re-defined so as to include events that occurred in what we usually classify as the remote past; and in the latter case, it should include events separated by temporal intervals of arbitrary size. Another option is represented by the so-called 'Alexandroff present' (see Arthur [2006], Savitt [2009] and, for a critical discussion, Dorato [2011]), i.e., the present identified with the spatio-temporal region corresponding to the intersection of the past light-cone of a later point and the future light-cone of an earlier point along the same 'world-line' (i.e., the path occupied in space–time by a persisting object). The good thing about this proposal is that the resulting region is both invariant and, if it is of the right 'size', truly able to account for the essential features of the experienced present – which, notice, is not point-like and has instead an extension, determined by our physiological apparatuses. The problem with it is, however, that it makes room for a degree of arbitrariness in what is to be included in, or excluded from, the present. How distant must the two points on the same world-line be in order for the present to be correctly represented? Even if we choose them so that their temporal distance coincides with the length of our actual experience of the present, what allows us to assume that the present is exactly the same for everybody?[35] Of course, one might simply claim that it isn't, and the Alexandroff present has to be shaped in different ways for different subjects. The problem with this is that, by giving priority to neurophysiology over physics, it sacrifices the objectivity of the present, which is however fundamental for presentists. Similar worries arise for those proposals that identify the present with a particular *conic structure* – one whose basis is centred on the observer and has a radius corresponding to the distance that a light signal moving towards us could travel before being experienced as successive to a given event in the immediate vicinity of us; and whose height is equal to such radius.

More generally, all relativity-inspired re-definitions of the present seem unsatisfactory for one fundamental reason. Namely, because the intuitive appeal of presentism consists in the fact that it promises to ground the belief that there is a unique, objective plane of simultaneity that is valid for every observer and that, consequently, every corner of the universe shares the same past, the same present and the same future. Whatever way one chooses of redefining the present in a relativity-friendly fashion, these promises are not fulfilled, exactly because of what appear to be the fundamental features of the physical theory itself. In this sense, doing our best to provide a physical basis for time as we make experience of it is *in principle* not enough.

In light of the foregoing considerations, then, it looks as though special relativity entails the rejection of presentism, lending support to the view that 'now' is an indexical, and temporal becoming just consists in a succession of events ordered in a four-dimensional block in which, when looked at 'from the outside', as it were, no point has a different ontological status from the others.

However, it cannot be concluded yet that the naturalistic metaphysician should be an eternalist. For, there is more to be said about the issue, and other relevant elements have to be considered. Since a lot has been put on the table already, though, any further discussion will be postponed to the next section. There, the two dichotomies introduced and analysed so far in this chapter will be further evaluated in light of the criteria for theory-choice suggested in Chapter 2; while doing so, other possibilities open to the scientifically-minded presentist will be critically examined.

2.   Metaphysics and contemporary physics of space and time: the verdict of the constructive naturalist

Summarising our discussion in the previous section, there are two pairs of opposing views that (naturalistic) metaphysicians have to look at when inquiring into the nature of space and time: one between substantivalism and relationism, and one between (A-theoretic) presentism and eternalism. Now, as far as the first pair is concerned, both substantivalism and relationism appear viable from the logico-metaphysical point of view. With respect to their fit with physics, substantivalism seems to be in a better position, but relationism is not completely ruled out, as – at least when applied to the temporal dimension – it meshes well with some recent new approaches to physical theory. In particular, it meshes well with at least one of the approaches that have been put forward with a view to eventually managing to unify relativity and quantum theory into a theory of quantum gravity, i.e., Barbour's. On the other hand, as we will see in a moment, some issues remain open concerning the capability of relationism to account for temporal experience and to avoid requiring one to dispense with time altogether. As for the second pair of concepts, coming up with a satisfactory definition of presentism is already hard, but appears possible. In that case, too, at any rate, no clear winner seems to emerge from merely *a priori* considerations (at least insofar as the presentist's ontological burden is not made too heavy and, for instance, a not-too-demanding truthmaking theory is preferred over a commitment to abstract truth-makers). When it comes to actual empirical data and scientific theories, however, presentism seems to be in bad shape, as the special theory of relativity appears incompatible with our intuitive notion of the present. However, as hinted at towards the end of the previous section, there are more elements that have to be brought to bear on the critical assessment of the presentism vs. eternalism debate.

Albeit with some disanalogies, then, we seem to be in a position whereby *a priori* analysis legitimates the reference to four different theoretical options; and looking at actual physics gives the edge to a four-dimensional substantivalist and eternalist view – where space and time form a unitary whole, all points of which are ontologically on a par. This is indeed the common opinion in the philosophy of physics community. At the same time, though, it looks as though more needs to be said before making this verdict final. Let us then do this now, looking at each pair of views in turn.

2.1   Relationism and antirealism about time

Several philosophers/physicists have recently argued that the latest developments in physical theory suggest that time is only an illusion (for an overview, see Callender [2010]). This revitalises certain ideas that have been already entertained in the past, most notably by McTaggart and Gödel. Let us briefly look at these older arguments and then move on to the more recent developments.

2.1.1   McTaggart and Gödel

McTaggart (see his [1908] and [1927]) can be interpreted as reasoning as follows:

a)   Time essentially involves change;
b)   Change amounts to something future becoming present and/or something present becoming past;
c)   Hence, regardless of whether being future, being present and being past are regarded as monadic properties or relations, change necessarily involves either contradiction (e.g., something is incoherently both future and present, both present and past), or circularity/infinite regress (different temporal statuses are attributed to the relevant entities at different times, so presupposing what was to be analysed, or at least a 'higher-order' something isomorphic to it);
d)   Therefore, change is not real;
e)   Therefore, time is not real.

McTaggart does not provide an argument in favour of (a) and just appeals to intuition there. The premise can be given support by emphasising the alleged impossibility of 'temporal vacua' (more on which later), but presentists certainly have the option of resisting it. As for, (b), B-theorists can (as first suggested by Russell) opt for a different definition of change: one in which events only occupy one fixed position in time, and thus never acquire a new 'temporal status'; and the fact that certain qualitatively different events are ordered in an objective earlier/later than sequence suffices for genuine change. However, on the one hand – at least according to McTaggart, who replied explicitly to Russell in the later reformulation (1927) of his original 1908 argument – this does not in any way circumvent the problem in (c), as it presupposes objective distinctions between past, present and future. On the other hand, the

Russellian strategy is of no avail to the presentist, at least to the extent that he or she is a 3D-, A-theorist and consequently rejects the idea of an objective ordering of temporal instants/object-stages which are all onto-logically on a par. Presentists, though, have an answer at their disposal: first of all, if future things do not exist, it cannot be the case that some-thing future becomes (also) present (and analogously for present objects/events becoming (also) past); secondly, if only present things exist, being past and being future cannot be genuine properties of concreta, let alone the properties relevant for an analysis of how concreta change; third, if the above is correct, change is correctly described as (numerically) the same concrete entities exemplifying (qualitatively) different prop-erties at distinct (present) times – that is, as the ever-changing present sequentially containing numerically identical but qualitatively dissim-ilar entities.

As for Gödel, he [1949] showed that there are solutions of Einstein's field equations in general relativity whereby a global time function cannot be defined; and took this to imply that time is not real (and is instead ideal, i.e., dependent on the mind). In more detail, Gödel proved the existence of cosmological models in which the universe is static and rotating, and there are closed trajectories in space–time, seemingly allowing for time travel. This latter fact means that, for any event in such models, that event is – among other things – earlier than itself; consequently, it is impossible to define a global time func-tion with the required features (there is no 'cosmic time'), as the past/present/future distinction is, at best, local. From all this, Gödel concludes that time cannot be real in these models and thus must be ideal *tout court*.

Einstein openly acknowledged the importance of Gödel's work for the development and understanding of relativity. However, it should be clear that the argument is far from straightforward and uncontroversial. Those who intend to resist it can argue in one ways:

1) They can claim that Gödel assumes that the lack of cosmic time makes time ideal, but one can deny that the reality of time requires its being cosmic (this appears problematic for presentists, though);

2) They can point out that Gödel assumes that the existence of accept-able models in which there is no cosmic time makes time ideal in general; but one can deny that the reality of time in *our* universe requires its *necessarily* being cosmic; in other words, the epistemic indistinguishability of our universe from a Gödel universe could be

deemed insufficient for accepting that we live in a universe without time.

In conclusion, interesting as they may be, both McTaggart's argument and Gödel's are not conclusive. Let us then move to the more recent considerations in favour of the unreality of time.

2.1.2 Contemporary antirealism about time

To begin with, it is undeniable that if one looks at physical theories, one can see that time does not in any obvious way play in them (all) the roles we naturally take it to play. Already in the equations of classical mechanics, for instance, time appears as a variable, but the flow of time in a particular direction is neither presupposed nor explicitly expressed. And the same goes for the 'now' that is taken to refer to the supposedly privileged instant of time that separates what is not real any longer from what is yet to happen: it simply plays no role in physical theory. Indeed, the idea that at least some fundamental problems affecting contemporary physics can be solved by eliminating time from the picture altogether is gaining popularity. And this means that getting rid of time might be not only possible but in fact necessary, and for reasons entirely internal to physics. Carlo Rovelli [2009], for instance, argues that as one moves to a deeper and deeper level of reality, time as a fundamental quantity disappears: physical theory turns out to be a theory of relations between variables, not a description of how things evolve in time. The t variable, says Rovelli, turns out to be nothing but a useful instrument, as time only emerges as a 'subordinate' entity at the thermodynamic level. Rovelli takes the elimination of time as a fundamental feature of any workable theory of quantum gravity, i.e., any theory that successfully merges general relativity and quantum theory.

Although this fact was not stressed in the earlier presentation of Barbour's ideas, his work is also paradigmatic in this sense. For, it is true that Barbour' primary aim is to defend the Machian conception of physical theory, which he regards as maximally simple and minimally committing. And his claim that time is unreal is – as we will see in some detail shortly – in a sense just the 'by-product' of Barbour's peculiar way of pursuing this objective. However, Barbour also states explicitly that the timelessness of his view of reality is crucial for putting general relativity and quantum mechanics together successfully – as he clearly believes to have done.

But if this is so, and that 'there is no time' is not only true in the sense that physical entities are not contained in an absolute (spatial and) temporal 'box' but, more importantly, in the sense that time is simply not an element of reality, then Barbour's Machian relationism is not something that relationists about (space-)time can exploit. Rather, not only does it not lend support to relationism about time in the traditional metaphysical sense, it also appears to dissolve the substantivalism versus relationism debate altogether – at least as far as time is concerned. Can anything more be added?

At this point, Barbour's account of temporal experience becomes relevant. We have mentioned Barbour's suggestion that the quantum probability distribution determines which configurations are most likely to be connected to others by 'minimal distance' paths across the space of Nows (Barbour's 'Platonia'). Crucially, this has to be understood in the sense that:

a) Certain configurations are 'more real' than others, and much more likely to be portions of reality inhabited by human observers making experiences;
b) (Exactly) these configurations are internally structured so as to contain 'time capsules': i.e., physical subsystems encoding information that our brains process *as if* it were information about Newtonian trajectories across canonical time.

If one follows Barbour's proposal, then, it looks as though one should be an error theorist about time: whenever we have the perception that something shifted from being possible to being actual, and from being present to having been present, we are in fact elaborating in our brains peculiar bits of information that are non-temporally 'written' in the physical configuration(s) in which we happen to find ourselves. These physical system are, basically, like static collections of movie frames. (As mentioned earlier, Barbour's universe looks like a completed n-dimensional puzzle,[36] the pieces of which correspond to possible configurations of the collection of all physical systems – recall his endorsement of a modified instants' Everettian interpretation of quantum mechanics). The question to be asked is, of course, whether this works. Baron, Evans and Miller [2010; 53] argue that it does not, and Barbour's account of experienced time is not satisfactory. One of their claims is that our perception of time might well be illusory, but it has to have certain 'dynamic' features that Barbour's picture is not able to reconstruct. Not only do we have to account for our distinctions between past, present and future events but also, and perhaps more importantly, for our changing perception and categorisation of those events and, consequently, for the mutability of the distinction itself. And Barbour (even leaving alone the increase in complexity when one moves to the intersubjective level, which Barbour does not discuss) doesn't provide such an account. Moreover, and even more importantly, it must be pointed out that Barbour never formulates a truly convincing argument in support of his claim that (some) solutions to the Wheeler–DeWitt equation give high probability to configurations containing time capsules – which is clearly crucial for his peculiar defence of an unyielding psycho-physical parallelism. At most, he offers a generic suggestion to the effect that the structure of the universe, together with the quantum 'probability mist', invariably determines the emergence of well-defined, unique and 'temporally well-behaved' histories from the multitude of possibilities contained in the quantum wavefunction.[37] But this is hardly sufficient.

But now consider: if it is true that Barbour's error-theoretic view of time is in fact incapable of fully recovering all the crucial features of time as we make experience of it from the allegedly timeless physical world and may, in addition, even lack a clear physical foundation, is it not advisable for the relationist about time to try to preserve the Machian perspective *without* also following Barbour in his antirealist views? After all, reductionism about *x* is not the same as, and should be kept distinct from, eliminativism about *x*, and it may plausibly be contended that Barbour (and perhaps other contemporary supporters of the 'time is an illusion' view) opt for the stronger eliminative thesis with no real reason for endorsing anything more than the non-eliminative, reductionist one. One may, of course, worry that Machian relationism, hence the formulation of a satisfactory theory of quantum gravity, and antirealism about time go hand in hand and, consequently, if one falls the other does too. However, this is not the case. While it is true that the route followed by Wheeler and DeWitt for quantising general relativity with specific initial constraints leads to a manifestly time-independent equation, this does not make eliminativism about time inevitable or highly natural. One might follow another route for quantisation, not requiring the Wheeler-DeWitt equation. And one might, more to the point given the role played by that equation in Barbour's theory, continue to use the Wheeler–DeWitt equation, but identify time as a function of the variables appearing in that equation.[38] Indeed, since all the rest of the technical setup remains unchanged, it looks as though one can certainly obtain the results Barbour is after without following him in his antirealism about time. (After all, whether paths across configurations give rise to an objective or merely perceived temporal structure cannot be

crucial for one's Machian solution to the critical issues besetting physical theory based on the postulation of those paths.)

Suppose, then, that the relationist identifies time as the sequence of
relations between distinct configurations of the universe, and regards
Barbour's 'best matching' relation as the direct expression of the fundamental nature of these inter-configuration connections. Does this give
us time as we intuitively conceive of it? Things are not so simple. As a
matter of fact, it is open to debate whether the alleged 'Parmenidean'
consequences of quantum gravity can be circumvented, and real time be
re-established so easily in the present context.[39] One way of putting the
problem is by asking whether, once the timeless universe gets 'unfrozen',
one can find enough structure in the theory to identify something that
deserves to be called 'time', in particular, something close enough to the
global dimension that we usually have in mind when we think about
time. Work on this issue is ongoing, and here we will just make two
remarks. First, strictly speaking, it would already be sufficient for relationism about time if one were able to identify various 'local times' with
the canonical sequential structure – for sure, accepting this is less revisionary than endorsing antirealism about time. Second, those aiming to
reconstruct a global time function – for example, because of a commitment to traditional presentist presuppositions – also have weapons
at their disposal: Gryb and Thebault [2012], for instance, show how,
contrary to the widespread conviction to the contrary, global time can,
in fact, be obtained by following a peculiar (admittedly non-standard)
quantisation procedure in contexts governed by a relational notion
of time.[40] We thus conclude that it is possible, and perhaps advisable,
to separate the new relationism championed by Barbour from antirealism about time; and that this is a potential basis for a naturalistically
respectable alternative to the canonical view of the metaphysical status
of (space-)time.

Before moving on to the second dichotomy (presentism/eternalism),
let us now consider first another issue which is directly relevant for an
evaluation of relationism about time, and which we left open earlier,
when discussing McTaggart's argument for the unreality of time: the
issue of temporal vacua. We have seen that relationism reduces time to
relations between physical entities. This is normally glossed as the view
that, according to relationists, time corresponds to change. As argued
by Shoemaker [1969], however, scenarios are conceivable in which not
only is the passage of time without any sort of change conceptually and
nomologically admissible, but this is also to be deemed probable given
the evidence available there; and this appears sufficient for putting

relationism into serious doubt. In more detail, Shoemaker (see also LePoidevin [1991; 94–98]) considered a world consisting of three disjoint regions, each one of which completely 'freezes' and remains changeless for a precise period of time at regular intervals (also becoming physically disconnected from the other two). These intervals are different for the three regions, so as to entail

a) That freezes in each region can be observed (or, better, indirectly reconstructed based on the available evidence) by the inhabitants of the other regions, who can then inform the inhabitants of the relevant region of what happened to their part of the world;
b) That some freezes can and do occur simultaneously in the three regions.

Shoemaker argues that (a) and (b) together make it reasonable to accept global freezes on the basis of the available evidence. However, global freezes differ from local freezes only in terms of extension, and since local freezes do not imply that time stops, argues Shoemaker, the same holds for global freezes. Therefore, the conclusion goes, relationism fails.

Against this objection, some (among others, Newton-Smith [1980; 42–47] and Butterfield [1984]) suggested a reformulation of relationism in modal terms (so that time is said to pass between two distinct instants t_a and t_b if and only if there is either an actual *or a possible* event occurring at an instant t_n in between t_a and t_b). Others argued instead that, upon scrutiny, it turns out that in fact "we are unable to conceive of a world about which it is clearly reasonable to claim that time passes but no events occur" and, thus, despite appearances to the contrary, Shoemaker fails to prove his point, for it is relationism itself that appears systematically more plausible (Warmbröd [2004; 282]). Is also possible to claim that change is, in fact, not required for real passage of time in the relationist framework, as sequences of *merely numerically distinct events* are sufficient. This reaction has not been explored in the literature and yet appears well-motivated. After all, events do not analytically entail qualitative novelty,[41] and if time is reducible to relations, why should it matter whether or not the relata of such relations are qualitatively distinct? If this is correct, the relationist can contend, contra Shoemaker, that time can, in fact, pass without change. When this happens, he or she will add, for every thing that appears to persist 'frozen' in the relevant interval, there is, in fact, a sequence of several events that are exactly similar qualitatively and yet non-identical (one may speak of absolute change/becoming, without qualitative change/becoming). Be this as it may, temporal vacua do not in any case seem too troublesome for relationists after all.

Another relevant theme, discussing which usefully allows us to link the issue currently under discussion to that having to do with presentism as opposed to eternalism, concerns the relevance of relationism (if any) for the dispute between A-theorists and B-theorists. It might be thought that, since relationism reduces time to relations, then realism about time understood relationally entails realism about the relevant relata, i.e., ultimately, the successive stages of objects exemplifying properties, and thus to eternalism. However, a presentist variant of relationism is conceivable that exploits the abovementioned presentist treatment of cross-temporal relations as relations (also) involving abstract relata. In particular, the presentist relationist can (in fact, must) argue that what is fundamental in his or her ontology is the (possibly) continuous existence of a three-dimensional 'slice' of individual things, constantly caused by their own being – together with that of the rest of the universe and the relevant laws of nature – to persist in time (or stop existing). In this context, persistence/continuous existence becomes fundamental, and, when regarded as a relation,[42] it always relates a specific individual to itself, that is, a concrete entity a to an abstract entity b that coincides with one of a's past or possible future states.

With this, we can now move on to a further assessment of the other debate that interests us in this chapter.

2.2   Presentism and eternalism: beyond relativity

We have seen in the previous section of this chapter that special relativity, no doubt a very empirically successful physical theory, seems to entail that we must abandon our intuitive view of the present and of time flow. Some of the considerations just made are already sufficient for seeing where more presentist-friendly alternatives might be found. But there is even more to say about presentism in the light of contemporary physics.

One first option for the presentist who does not want to simply drop special relativity in order to preserve his or her philosophical beliefs is to revise the established understanding of the theory. Strictly speaking, the latter is the by-product of Minkowski's [1907] geometrical interpretation of Einstein's kinematics as originally presented in 1905. But the latter was (or, at least, can be regarded as) a mere 'theory of principle' (aiming, that is, to account for the phenomena) rather than a 'constructive theory' (aiming, that is, to describe an underlying reality). Moreover there is a constructive view that avoids regarding the lack of a privileged frame of reference as an objective feature of reality. It is the so-called 'neo-Lorentzian' interpretation of special relativity, in which length contraction and time dilation *within a privileged frame of reference* (playing the role of the immobile ether postulated by physicists until the end of the 19th century, and which Lorentz tried to preserve in spite of peculiar experimental results that led to Einstein's revolution)[43] is postulated so as to account for the evidence in an essentially classical context. Crucially, the neo-Lorentzian interpretation of special relativity allows one to leave the formalism of the theory untouched and is therefore empirically equivalent to the more popular alternative. And, of course, presentists can exploit the existence of a privileged frame of reference in such an interpretation with a view to claiming that their metaphysics is, after all, compatible with, if not directly supported by, physical theory. Indeed, authors such as Craig [2001] and Bourne [2006] explored this option in some detail. Among other things, they stress that the customary relativistic definition of simultaneity presupposes that light moves at the same speed in one direction across space–time and in the reverse direction, but this is by no means obvious, as it crucially depends on certain homogeneity and isotropy conditions that are normally simply assumed to be satisfied by space–time. Remarkably, it is also argued that the neo-Lorentzian interpretation might also enable one to solve the problem of providing a relativistically invariant formulation of quantum mechanics.[44] Setting aside more specific problems besetting the neo-Lorentzian interpretation of relativity,[45] however, two fundamental worries arise on such construal:

a) That Lorentz invariance appears to be merely accidentally shared by the laws that effectively govern physical systems in Newtonian space–time (see Balashov and Janssen [2003; 341]);
b) That the privileged frame of reference remains, in any case, a merely postulated unobservable entity.

Generally speaking, the idea that the fundamental structure of space–time is, in fact, the origin of both the universality of Lorentz invariance and the impossibility to detect a privileged frame appears compelling, and to consequently favour the established interpretation of special relativity over the neo-Lorentzian interpretation.

Based on the fact that special relativity is not the most fundamental theory presently available, however, the presentist may hope to find better support in other theories. In particular, since special relativity is obviously less fundamental than general relativity, perhaps the latter is the theory that presentists should look at, especially in view of the possibility that something like an absolute time might be found at the level of cosmology. The idea has been explored, in particular, of reconstructing the much-needed privileged frame of reference by looking at available solutions to Einstein's equations that coincide with models of the universe with the appropriate features. This is tantamount to identifying Big Bang models in which the expanding universe is homogeneous and isotropic, that is, has (more or less) the same values of density and pressure in all regions. Now, such models exist and are those described by the so-called 'Robertson–Walker metric'.[46] Hence, it might look like presentists can pursue this strategy successfully and suggest that our universe actually has the features described by the Robertson–Walker metric. However, worries have been raised (Callender [1997], Bourne [2006]) concerning both the technical workability of this proposal and its general methodological plausibility.

Some authors have pointed out that quantum mechanics presupposes an absolute time. But this is, in itself, of no avail to presentists, as they should additionally argue that quantum mechanics is somehow more fundamental than relativity – in particular, as a theory of (space–time). Here is where we are led back to quantum gravity – the theory, or family of theories, expected to incorporate both the principles of general relativity and those of quantum theory that we have already introduced and discussed in earlier sections. Monton [2006], for instance, notices that there are approaches in so-called 'fixed-foliation quantum gravity', such as those relying on foliations of space–time into hypersurfaces of constant mean (extrinsic) curvature, which are compatible with presentism. Of course, it is questionable whether these are the versions of the theory that one should prefer, and thus whether this suffices for re-establishing the priority of presentism over eternalism. Still, if Monton were correct, it would at least be the case that those having independent reasons for believing in presentism might insist that the approaches to quantum gravity in question are indeed those to be pursued. Wüthrich [2010a], however, makes a forceful case for the existence of internal weaknesses in the approaches to quantum gravity that Monton has in mind; and also for doubting that such approaches truly lend support to presentism anyway.[47]

At this point, since all other alternatives (at least those considered here) fail to provide solid grounds for presentism, perhaps the more radical choice of abandoning the basic relativistic tenet that space and time constitute an indivisible four-dimensional whole might represent a more fruitful choice for the presentist. Of course, this is tantamount to endorsing the sort of new relationism *à la* Barbour without antirealism about time that we have looked at earlier in this chapter. The pros and cons of Barbour-like, non-4D, relationist approaches to quantum gravity have already been discussed and, while there certainly is more work to be done, it seems fair to suggest that these represent at least an interesting avenue for future research.

2.3 Taking stock

It is now time to suggest an overall critical assessment, based on the criteria for theory-choice identified in Chapter 2, of the issues surrounding space and time that have been discussed in this chapter.

Starting from the substantivalism vs. relationism opposition, the two views appear to be equally general and open to a confrontation with our best physical theories. As for simplicity, relationism – at least once Barbour's Platonia view, with all configurations enjoying the ontological status, is dropped – fares better in terms of ontological economy. (At least one fundamental entity, time, is analysed in terms of other entities.) But substantivalism requires less complexity in terms of how physical theory is formulated and reconstructed. (Recall, for instance, Barbour's need to use conformal superspace to avoid the problem of indeterminism.) Therefore, even though Machians claim to be able to 'do physics with fewer parameters', no clear advantage for one view over the other can be found here. Now for the crucial criteria. In terms of being close to common sense, it doesn't seem that either substantivalism or relationism is better off per se. However, we have suggested, the relationist stance based on a '3 dimensions+1' approach to physics might be the basis for a form of presentism, and could consequently share the clear advantage enjoyed in terms of conservativeness by presentism and by A-theories of time more generally. Additionally, in terms of fruitfulness, relationism might be able to break the tie in its favour, provided that it really proves able to satisfactorily unify distinct theoretical domains in the way that authors like Barbour claim. At the same time, though, as far as empirical adequacy is concerned, substantivalism still seems to deserve the preference it has been accorded in the last 50 years or so. For, it is widely agreed that substantivalism offers a better account of the nature of space–time as it is described in general relativity, and a better basis for doing physics more generally. In view of the foregoing, a fair, the necessary conclusion to draw for the time being is that substantivalism is still the option to favour. However, relationism must be taken seriously and might even reverse the order of preferability if it really proves theoretically useful as it is presented to be by its contemporary supporters, solves its outstanding technical problems and manages to define a realist stance with respect to time (perhaps also successfully combining with the A-theoretical perspective on the nature of time and becoming).

Looking now at the presentism vs. eternalism contrast, here, too, the two alternatives appear to be on a par in terms of generality and refutability. In terms of simplicity, things are not really straightforward, but it seems plausible to claim that while presentism is less committing in terms of existence (only the set of things that simultaneously exist presently exist at all),[48] eternalism proposes a simpler structure and a uniform account of the temporal dimension, and also has fewer problems to solve at the purely *a priori* level. Assuming that this amounts to a non-negligible advantage for eternalism, let us consider empirical adequacy next. As things stand now, it is simply a fact that presentism fares much worse than eternalism, as nothing in our best physics lends support to the idea of an absolute time and of objective, universal distinctions between past, present and future. True, conservativeness strongly advises us to choose presentism over eternalism. But in no way can the naturalistic metaphysicians draw his or her conclusions *exclusively* on the basis of the conservativeness criterion, unless the alternatives are essentially equivalent *with respect to the other criteria* – which is certainly not the case here. In light of this, it looks as though eternalism should be preferred to presentism. Here, too, however, there is an alternative that must be kept in mind: namely, that constituted by a form of presentism combined with a Barbour-type form of relationism about time only. (We have already offered a sketch of what such a presentist relationism would look like towards the end of Section 2.2.) Of course, when presented in that form, presentism becomes parasitic on the future developments of the relationist programme in the philosophy of physics. Were this latter programme successful, however, perhaps the presentist might find a more appropriate basis for insisting on the existence of a fundamental, privileged frame of reference coinciding with the present as he or she understands it. For instance, instead of trying to reconstruct the privileged frame of reference based on complex empirical assumptions about the universe as modelled by the (Friedman–Lemaître–) Robertson–Walker metric, he or she could insist on the separation of the three spatial dimensions from the temporal dimension and *consequently* read the Big Bang literally: that is, as the birth and evolution at successive instants of time of all the material things that exist (each one of the these things being simultaneous with every other at each instant, and only one instant being 'real').[49]

In this chapter, too, as in the previous one, the 'verdict' that has been presented is by no means intended to be a conclusive one (as a matter of fact, it was presented as a possible exemplification of a general methodology, and its details can certainly be put into question). Still, it is believed that at least some objective facts about certain metaphysical options and their relation to physical theories have emerged and have been usefully put into a broader picture. Interestingly, the implementation of the proposed methodology and criteria has led here (albeit with a number of provisos and alternatives left open) to conclusions that are much further away from common sense – hence, closer to physical theory in its more 'revisionary' philosophical interpretation – than in the previous case. Perhaps, this is a sign of the reasonableness of the methodology and criteria themselves?

Leaving this to readers to decide, we will now conclude this chapter by briefly looking at another theoretical option that needs to be taken into account in the context of a scientifically-informed philosophical analysis of space and time.

3. Supersubstantivalism

Suppose that substantivalism were indeed to be preferred to relationism. Then, a new option opens up for the metaphysician: namely, that of dispensing with the other fundamental kind of substance, namely material objects. How can this be done? By endorsing *supersubstantivalism*, the view that material objects are identical to spatio-temporal regions (exemplifying properties), and thus there exists only one substance, space–time. Strictly speaking, one might claim that objects *are* spatial regions and that the only existing substances are spatial points, leaving it open whether or not time is also a substance – by now well-known considerations about relationism in quantum gravity might be a good reasons for doing this. Here, however, we will only discuss space–time supersubstantivalism, i.e., the four-dimensionalist view according to which material objects are identical with space–time 'worms' of sorts, tied to properties. By and large, the considerations that will be made in the rest of this section will extend virtually unchanged to the 3D case.

On the one hand, supersubstantivalism has become increasingly popular among philosophers in recent years: it has been defended, for instance, by Sider [2001], Skow [2005] and Schaffer [2009a]. Among other things, it has been presented as more parsimonious than a substance dualism of spatio-temporal points and material objects; as able to naturally explain the fact that material objects have geometrical and mereological properties that mirror those of space–time regions and the fact that material objects cannot exist without occupying space–time regions; as being in harmony with General Relativity, where the distribution of matter is given by the stress–energy tensor, which defines a field and is, consequently, naturally interpreted as a property of space–time; and as also finding support, for similar reasons, in quantum field theory. As for the four-dimensionalist component, it also appears to bring advantages with itself: for instance, how is one to explain the motion of material objects, given that parts of space do not move around? In a three-dimensionalist supersubstantivalist setting, this would be a problem, but a four-dimensionalist supersubstantivalist can argue that motions can (and should) simply be reconstructed in terms of continuous space–time lines.

On the negative side, supersubstantivalism certainly is a radical metaphysical view, and one with a number of open issues. To mention one that has been discussed in the literature, the supersubstantivalist has to account for basic intuitions that seem to be lost if one endorses his or her views. For example, I could have been to the left of where I actually am, but the space–time region that I currently occupy is essentially where (i.e., the one) it is, hence – it could be argued – I cannot be a space-time region. Ways out of this have been identified: supersubstantivalists can explain away the relevant differences as appearances; or they can give up geometrical essentialism – the doctrine that points of space–time have their properties essentially; alternatively, they can abandon compositional essentialism for space–time regions – the view that the latter cannot contain different points in different possible worlds. (There seems to be no obvious way to proceed here, but see Skow [2005; paragraphs 3.7–3.8] for the suggestion that giving up geometrical essentialism is the best option on the basis of the key features of general relativity.) Another potential problem (which, unlike the previous one, seems to have gone virtually unnoticed so far) also exists, having to do with co-located numerically distinct objects (of the same kind): as we have seen in Chapter 3, standard quantum mechanics explicitly allows for the possibility of distinct physical systems having the same location – more specifically, the same probability of being found in a specific location, where probability assignments, however, have objective and not epistemic significance. But this means that supersubstantivalists who take quantum theory seriously (or at least understand standardly interpreted quantum mechanics as describing a possible world that they have to account for) will have to say something more about what properties can become tied to space–time regions and how. Perhaps the best option in this respect is to endorse the field-theoretic reconstruction which, as we have seen, does away with the idea of distinct co-located objects and replaces it with the notion of field-parts exemplifying properties. This is plausible on independent grounds, and also in harmony with some of the reasons motivating the endorsement of supersubstantivalism in the first place. (It remains a fact, in any case, that supersubstantivalism appears simply incompatible with (the standard interpretation of) non-relativistic quantum mechanics.)

Be this as it may, the key question with supersubstantivalism, at least from a naturalistic point of view, is obviously whether it truly finds support in contemporary physics. An assessment of whether matter fields are truly identical to space–time certainly requires more than a generic reference to general relativity and quantum field theory. If one restricts fundamental properties of things to geometric properties, one may employ geometrodynamics with a view to making supersubstantivalism gain support.[50] However, independently of the role it plays in theoretical frameworks such as Barbour's, the geometrodynamical programme is, as mentioned earlier, far from being a progressive one. Besides, more modest forms of supersubstantivalism, allowing for space–time points to exemplify more than merely geometrical properties, would be more plausible – if only because they would allow one to avoid rewriting all physical laws as laws governing the curvature of space–time. In this case, however, the extent to which physical theory renders supersubstantivalism plausible is even less clear, and much more work is to be done.[51]

For the time being, then, we will close our brief discussion of supersubstantivalism and just leave it there as an alternative to be assessed carefully by naturalistic metaphysicians in the future, perhaps on the basis of theoretical construals anlogous to Barbour's.

4. Conclusions

In this chapter, it has been argued that – given the current status of research in physics – there is reason for endorsing the most popular (at least in the scientifically-oriented context) philosophical views, according to which space and time form a unitary four-dimensional whole, within which the distinctions between past, present and future (hence, becoming) do not have an absolute, objective status. Given the unquestionable relevance of the commonsense intuitions at stake (at least in the case of A-theoretic presentism), and in view of the open possibility of new developments and alternative interpretations in physics, at the same time we suggested that two other alternatives are, in any case, worth taking seriously. On the one hand, a form of relationism about time inspired by work such as Barbour's (but in the context of which Barbour's antirealist stance with respect to time itself is dropped) – possibly connected to a specific form of presentism. On the other, a radical form of supersubstantivalism whereby objects do not occupy regions of space(-time) but rather are identical to them. If super-substantivalism is applied to space only, of course, the two metaphysical pictures can be put together so as to provide a simple ontology of substantial entities provided with geometrical and other properties, and whose shift from one configuration to another gives rise to the temporal dimension. At any rate, it would seem that, much more than in the previous case study, further research is needed, and new results to be auspicated if more is to be said about the nature of space and time. And here, too, any real philosophical progress is only likely to come from a truly naturalistic approach to metaphysics.

Having said this with respect to the nature of matter and of space and time, in the next chapter we will move on to a consideration of our last case study: it will concern the nature of the part–whole relation in all of its aspect – from composition, holism and supervenience to ontological dependence, fundamentality and 'Spinozean monism versus Democritean pluralism'.

Notes

1. One important issue that will not be discussed in this chapter is whether space and time need to be considered fundamental or are instead emergent from an underlying physical basis which is not spatio-temporal. Our analysis in what follows may seem to take the first option for granted, but this is not so. Indeed, the discussion in this chapter is, by and large, compatible with the view that the universe is not the spatio-temporal fusion of basic spatio-temporal parts. For an explicit defence of this latter claim (and of the alternative view of the universe as a mereological composition of properties), see Paul [2012a], where the spatio-temporal view is connected to the thesis of Humean Supervenience that we will discuss in Chapter 5.
2. The definition of these two positions is, as a matter of fact, quite controversial, as it is not easy to define them and distinguish them from each other in a clear-cut way. We will attempt to provide a less controversial description of this dichotomy in what follows.
3. In particular, Newton's views on time, space, place and motion are expressed in a 'Scholium' at the beginning of the *Principia*, inserted between the 'Definitions' and the 'Laws of Motion'.

4. Notice, however, that this does *not* entail that the predicate 'is in true motion' has no meaning in a Cartesian context. We will not get into a detailed discussion of this here.

5. Things are more complicated than this. The rotating bucket experiment is the last of five arguments 'from the properties, causes, and effects of motion', which were designed by Newton to show that an adequate analysis of true motion must involve reference to absolute space. Newton also offered the example of a pair of globes connected by a rope and revolving about their centre of gravity. This he took to show that, despite the fact that absolute space is invisible to the senses, it is nonetheless possible to infer the quantity of absolute motion of individual bodies. For details, see Rynasiewicz [2011].

6. Because change can be faster or slower, but time cannot, and change is in or where the changing thing is, whereas time is "alike everywhere and with all things" (218 b 9–18).

7. Which means that the first distribution is transformed into the second in a way that leaves the fields unchanged outside a particular region even though they are spread differently outside of it – the spreadings inside and outside being related in a uniform, non-abrupt manner. More technically, a hole transformation of a field M is a diffeomorphism on M that is the identity outside some arbitrarily selected neighbourhood but comes smoothly to differ from identity within that neighbourhood (see MacDonald [2001]).

8. It can be plausibly argued, however, that a four-dimensional space–time ontology allows the substantivalist to do away with absolute velocities while still solving the problem represented by kinematically shifted universes (see Pooley [forthcoming]). This advantage of relationism over substantivalism might, therefore, be more apparent than actual.

9. The initial starting point being the Jacobi theory of mechanics, in the context of which, says Barbour, time turns out to be a measure of duration which emerges entirely from the dynamics and does not pre-exist at the level of pure motion, considered independently of its causes and its changes.

10. Importantly, it can be argued (see Belot [1999]), that the one in question is not really an assumption, possibly *ad hoc* with respect to the aim of constructing a completely Machian physics, but rather a prediction entailed by Barbour's reconstruction of physical theory. And one, Belot suggests, that, as a matter of fact, appears to be empirically successful – in which case, of course, one may even regard relationism as providing the missing explanation of a seemingly contingent, but fundamental, physical fact.

11. In quantum mechanics, Schrödinger's equation describes how physical systems evolve in time. Usually, it is given in the time-dependent form

$$i\hbar \frac{\partial}{\partial t} \Psi = \hat{H} \Psi,$$ where Ψ is the wavefunction of the quantum system, i is the imaginary unit, $\hbar$ is the reduced Planck constant, and $\hat{H}$ is the Hamiltonian operator, expressing the total energy.

12. In connection to this, it can be argued (see, e.g., Pooley [forthcoming]) that Barbour's construal is still naturally committed to substantivalism, as the properties characterising physical configurations are best conceived of as possessed by point-like substances. More generally, it might be insisted with some degree of plausibility that the spatio-temporal manifold and its metric

are not fully analysable in relationist terms, as they are physically real in their own right. This is a complex issue that deserves careful assessment. Here, we will just point out that, even granting this need for substance, Barbour's anti-realism with respect to time seems to entail that such need can (and perhaps must) be limited to substantivalism about space. (Pooley does not seem to question the timelessness of Barbour's universe, at least not based on the idea just pointed at.)

13. Of course, strictly speaking, time series cannot be orderings of events but only of instants. It is possible, however, and certainly advisable given McTaggart's own use of these notions, to talk interchangeably of instants of time and of events occupying those instants at particular places (and, thus, of particular objects exemplifying properties at particular spatial and temporal locations).

14. Following McCall [1994], one may add to possibilism the view that the future consists of a number of branches, only one of which becomes actual at each instant (in which something happens that could have been otherwise). It is also possible – although it is not a widespread view in the philosophical literature – to reject the tenseless view by believing that the future is real and as time flows the total amount of existence decreases as things cease to be real by becoming past (see Casati and Torrengo [2011]). It must also be noted that a form of 'transient presentism' can be opposed to standard presentism, according to which the past exists but only as a set of atemporal facts (see Fiocco [2007]).

15. Notice that A-theorists do not endorse the Meinongian view according to which there are things that do not exist. According to them, there simply are no past/future events or objects intended as concrete spatio-temporal entities. (These theorists might, however, postulate abstract entities additional to presently existing concrete ones.)

16. From now on, we will restrict our focus to standard presentism, ignoring the growing block and the shrinking future A-theoretical views.

17. While Skow works in an eternalist setting, Brogaard rather speaks of four-dimensional objects, so making it possible that, in spite of the fact that not everything that exists is fully present, the overall framework is A-theoretic. For different arguments in favour of the claim that presentism need not be a version of the A-theory (based on the idea that A-properties may be reducible to B-relations, or even be dispensed altogether), see Rasmussen [2012] and Tallant [2012].

18. Meyer also considers the idea of something 'existing simpliciter', and rejects it based on the idea that to exist simpliciter can only mean to either exist temporally or exist outside time or exist in some other possible world; and, while the first conjunct just turned out not to be able to do the work that presentists want it to do, the other two are simply irrelevant.

19. For instance, Sider believes that presentists and eternalists agree that dinosaurs once existed, but eternalists will express this by saying something like 'There exist dinosaurs, located temporally before us', that is, in the formal terms of predicate logic, $\exists x(Dx \& Bxu)$, while presentists will instead claim that 'It was the case that: there exist dinosaurs', that is, $P\exists xDx$ (where 'P' stands for 'It was the case that').Thus, presentists and eternalists can agree on the truth value of the initial claim about dinosaurs in the past, but they disagree on its

logical form. However, says Sider, in evaluating ∃x(Dx&Bxu) *both* eternalists and presentists regard the existential quantifier as absolutely unrestricted, hence they disagree on something precise, and substantial enough: namely, the truth value of the entailed claim that there are dinosaurs, ∃x(Dx). The problem with this (as argued, for instance, by Tallant [forthcoming a]) is that to deny that P∃xDx entails ∃xDx, one has to either reject the claim that 'It was the case that there now exists a D' entails that 'There now exists a D', but everybody will agree on that rejection; or the claim that 'It was the case that there has been, there is now or will be a D' entails that 'There has been, there is now or will be a D' which, however, appears instead perfectly acceptable.

20. This has been questioned (e.g., by Keller and Nelson [2001] and Bourne [2006; 133]) based on the Lewisian idea that time-travel merely involves a discrepancy between 'personal' and 'external' time (together with an appropriate form of connectedness between the relevant 'stages' of the traveller); and that stories of this sort are compatible with presentist ontologies. It can plausibly be argued, however, that this merely means to construct an *ad hoc* notion of time travel and, more importantly, time. See, for instance, Hales [2010], who discusses forms of presentism committed to a two-dimensional (see Meiland [1974]) or, at any rate, multi-dimensional time – which seems to be the only thing that allows for some sort of time travel in a presentist setting. In general, it looks as though, instead of trying to show that their metaphysical view is compatible with time travel, presentists should use time travel (and a related claim of impossibility) as a tool for rendering their position legitimate. There is, however, no need to insist further on this point here. On time travel and physics more generally, see Arntzenius and Maudlin [2010] and Smeenk and Wüthrich [2011].

21. A recent elaboration of these ideas, which indeed improves significantly on previous formulations of the possible worlds/metaphysics of time analogy, can be found in Noonan [2013].

22. Another way of putting this is by saying that if presentism is true, then the falsity or truth of claims about the future is always provisional because determined only *indirectly* by states of affairs that are not identical with what would (or, will) count as the truth-maker for those claims if they were (or, once they become) claims about the present. The foregoing is also relevant for the so-called 'problem of future contingents'. Such problem was originally identified by Aristotle in chapter IX of his *On Interpretation*: any contingent claim about a future time *t* (e.g., that there will be a sea battle tomorrow) will come to have a definite truth value at *t*. But truth values would seem to be atemporally possessed, hence it must be concluded that the claim in question was already true/false before *t*. And since all past truths are necessary, it looks as though all claims about the future are necessary and the future, even though it might be not real yet, is completely determinate. Eternalists will not have too much of a problem with this, but not so for presentists. Besides having recourse to the above differentiation between direct and indirect truth-making, however, the latter can allow for the 'neither true nor false' truth value in the case of claims about the future.

23. Stoneham, too, notices that "the eternalist cannot deny bivalence for future contingents" [2009; 214], and argues that, provided that all parties agree on some form of the truth-maker principle (If a proposition *p* is true then there

is some object *x* such that the existence of *x* strictly implies the truth of *p*), then the debate between presentists and eternalists is genuine. Tallant [forthcoming a] complains that this requires truth-maker maximalism, the view that every true proposition, including those stating the non-existence of something, requires an actual truth-maker. He then goes on to notice that if one restricts one's truth-maker principle to propositions about what exists, then the ambiguity with existence again represents a problem. This is correct, but (a) presentists can, of course, insist on truth-maker maximalism, perhaps accepting a commitment to negative facts; and, in any event, (b) the key point being made here is not one concerning one's theory of truth but rather the existence of genuine differences between two metaphysical frameworks.

24. The question is, then, whether or not moving at the level of the most general categories suffices for individuating a sense of existence different from, and more fundamental than, those aspects of it mirrored by our – necessarily tensed – language and thought. Perhaps Tallant's strategy is just a new variant of the 'existing simpliciter' strategy already mentioned, and dismissed, earlier. Perhaps not.

25. Other examples include causal relations and relations whereby past entities constitute the contents of present thoughts or the referents of presently uttered words.

26. Cameron rejects the idea of properties such as 'being such that it was the case that...' based on the fact that such properties do not encode anything about the present intrinsic nature of presently existing things. Temporal distributional properties allow one to steer clear of this difficulty, as they are not only presently exemplified but also characterise (among other things) the present intrinsic state of their bearers. For objections to Cameron's construal, see Tallant and Ingram [2012].

27. This is the basis for what Bourne calls 'ersatzer presentism' [2006; 52–70]. See also Wüthrich [2010]. For a general discussion of the various options available to presentists with respect to the grounding problem, see Davidson [forthcoming].

28. This means that "while *that there were dinosaurs* doesn't depend for its truth on how the world is, it does depend for its truth on how the world was. The presentist can then hold that while *that there were dinosaurs* lacks a truth-maker, *that there are dinosaurs* used to have a truth-maker. On P-Grounding [the proposed account of the grounding of truth], this is sufficient for treating *that there were dinosaurs* as grounded" [Ib.; 346]. Of course all requests for 'second-order' truth-makers need to be rejected.

29. In other words, while presentists are said to be in trouble in accounting for things that it is necessary for any theory of time to account for, eternalists might be happy to maintain that their theory is to be preferred based on philosophical analysis (and, perhaps, scientific data – more on this in a moment) *even though* this forces us to reconsider entrenched beliefs (in particular, our, perhaps naïve, views about moral responsibility and free will).

30. That this is more than just a possibility can be argued for on the basis of the quantum correlations exhibited in entangled systems discussed in Chapter 2. Rietdijk [1966] and Putnam [1967] first exploited the existence of directly correlated, hence equally real, measurement results at distances that light

cannot travel instantaneously to show that the future/present/past distinction cannot be frame-independent.

31. Much of what follows is drawn from Dorato [2013]. Also see Savitt [2000].

32. Skow suggests that the theory might include elements from the growing block view, so re-establishing an ontological asymmetry between the future and the present (and the past). However, the resulting view only seems to enable one to talk of regions of space–time that do not contain any events *from the point of view of specific observers*; hence, again to fall short of providing a truly satisfactory surrogate for what we really look for, that is, an *absolute* present and a truly open future.

33. See Godfrey-Smith [1979] and Hinchliff [2000].

34. Weingard [1972].

35. As McTaggart already argued, it most certainly is not, as human perception differs from individual to individual, most radically so in pathological cases – for instances, those known ones in which the present extends enormously, and the individual has almost no perception of time passage and merges in a unique whole experiences he or she had weeks, months or even years apart from one other.

36. Where n denotes the number of relative quantities that need to be employed to provide a complete description of each instantaneous configuration of the universe.

37. This is based on work done in the late 1920s by Heisenberg and British physicist Nevill F. Mott with a view to explaining the emergence of linear tracks in cloud chambers in spite of the sphericity of the pre-observation wavefunctions.

38. Recall Rovelli's [2009] views, mentioned – albeit in passing – a few paragraph above.

39. Earman [2002] is pessimistic, but see Maudlin [2002] for the contrary opinion.

40. In particular, they argue that standard canonical quantisation procedures are generally unsatisfactory, and also responsible for the problem of time; and that, once General Relativity is translated into the formalism of the above-mentioned shape dynamics, *modulo* certain formal extensions of the theory standard quantisation can be implemented so that it leads to a dynamical theory of quantum gravity which retains a canonical temporal structure (while *not* reintroducing an absolute, non-relational notion of duration).

41. Indeed, it has been argued that primitive identity should be attributed to events (Diekemper [2009]). This also follows, of course, on a minimal conception of an event as the exemplification of a property by an object at a time. Indeed, relationists could, more directly, stop assuming an ontology of events and revert to one of primitively individuated objects. In fact, traditionally relationism about time was indeed formulated in terms of objects: Leibniz, for instance, said that "instants... consist only in the successive order of things" [1704(1956); third paper, section 6]. And while an ontology of events appears natural in the relativistic context, this is no longer the case in the context of the 'new Machianism', as it separates, more or less sharply, space from time. Indeed, as we have seen, Barbour explicitly talks of configurations of objects in space. In view of this, as already suggested in the context of the discussion of the ontology of quantum fields in Chapter

3, an ontology of objects and not events might be preferred because closer to common sense. Of course, all this needs to be articulated further. But there are ways of doing so. Since Russell's [1914] claim that time instants are to be construed as maximal sets of pairwise simultaneous events, various formal treatments have been offered that substantiate this fundamental claim (see, for instance, Pianesi and Varzi [1996], where temporal relations are reduced to mereo-topological properties of events). But Chisholm [1990], for instance, provided a relationist treatment of time whereby the fundamental notions employed by Russell (e.g., 'wholly precedes', 'overlaps') are systematically turned into relations involving objects and properties and nothing else.

42. Especially from an endurantist point of view (according to which things persist in time by being wholly present at each stage of their existence, and not – as perdurantists have it – by having distinct temporal parts), persistence would really seem to be a monadic, intrinsic property grounding certain temporal relations, not the converse. But this, and the entire endurantism/perdurantism debate, are something we do not need to spend time on in the present context. (Another set of considerations that seem to deserve scrutiny, but cannot be dealt with here, concerns the details of the ontological pictures that emerge from the various combinations of substantivalism/relationism about space and/or time on the one hand, and A- or B-theoretic views of time on the other).

43. In 1887, Albert Michelson and Edward Morley attempted to detect the relative motion of matter through the stationary luminiferous ether, but nothing of the sort was observed. This is normally taken to be an effective refutation of the pre-relativistic view of the physical world.

44. In particular, of the collapse of the wavefunction or of the dynamics obeyed by wavefunctions in no-collapse interpretations.

45. Discussed, for instance, by Balashov and Janssen [2003].

46. More precisely, the Friedman–Lemaître–Robertson–Walker metric, a special sort of non-Euclidean metric that describes a homogeneous, isotropic, either simply or multiply connected (each point in space–time being represented by either one or more than one set of coordinates) and either expanding or contracting universe.

47. Wüthrich also adds that what he regards as the most promising approaches to quantum gravity (that is, string theory and loop quantum gravity) make no room for a privileged frame of reference. Witten's interpretation of string theory, does reconstruct time as a fundamental magnitude based on the metric defined over string-points that, together with a particular multi-dimensional field, he takes to constitute the theory's ontology. However (see Huggett, Vistarini and Wüthrich [forthcoming]), on such a construal durations seem to lose physical significance, and thus the extent to which one truly has 'found time' is unclear to say the least.

48. Tallant [forthcoming b] argues that presentism fares better than eternalism in terms of quantitative economy: in particular, the presentist posits fewer entities than the eternalism *and* does not owe us an explanation that the eternalist owes us – namely, of why we engage in 'stubborn quantifier restriction', refusing to quantify unrestrictedly over entities past, present and future even though we (should) believe that they are all on a par.

49. Notice, incidentally, that on a B-theoretic reading, the Big Bang is really just a claim about the shape of a four-dimensional whole that does not change – simply for the fact that it contains time within itself.
50. See Misner and Wheeler [1957]. The contemporary re-elaboration of the Cartesian idea that matter is pure extension and no property-bearers distinct from space–time points need to be postulated dates back to the work of Herman Weyl (see his [1927(1949)]) and of William Kingdon Clifford before him ([1870]). Clearly, there is an interesting connection here with our earlier discussion of relationism.
51. Relevant work in this sense has been done recently: for instance, Lehmkuhl (see his [2011]), argues that mass, energy and momentum are extrinsic properties of objects that depend on space–time; and that matter depends essentially on space–time, but not vice versa. This might justify the move to a radical form of monistic substantivalism but, as Lehmkuhl himself notices, whether – and, if so, to what extent – general relativity and/or unified field theories allow for/push towards supersubstantivalism still requires careful study. Also worth mentioning in this context is Wallace and Timpson's [2010] 'space-time state realism', according to which the states associated to space–time regions are fundamental – a view that these authors take to bring a number of theoretical advantages with itself, but do not consider as necessarily a form of supersubstantivalism.

5
Parts and Wholes

1. Part–whole: definitions and issues

An important set of questions in analytic metaphysics, and one which is often taken to lack real significance and interest from the empiricist/naturalist viewpoint, concerns mereology and its application to the inquiry into the nature of reality. Recall, for example, the question that Putnam takes to be paradigmatic of the uselessness of metaphysics, and that we discussed in Chapter 1: "How many objects are there in a three-particle universe?" If a metaphysical question is to have any significance at all from the naturalistic perspective, it must be somehow connected, at least potentially, to empirical data and scientific theorising. But how can this be done, when all we have to work with are abstract principles endowed with merely formal significance?

The sections to follow will not attempt to provide a general defence of the metaphysical import of mereology from a naturalistic viewpoint. Nor, for that matter, will they introduce mereology and its principles in any detail. Rather, we will look at some open issues in the philosophy of physics and at some ongoing debates in metaphysics that appear to share a common conceptual ground, based, among other things, exactly on various aspects of the part–whole relation. It will turn out that in this case, too, as in those discussed in the previous two studies, there is a possibly fruitful interplay between *a priori* philosophical analysis and the *a posteriori* considerations coming from science.

In more detail, the rest of the present chapter will deal with two general issues related to parts and wholes – and then apply the criteria for theory-choice proposed earlier to the options that will be identified in the course of the discussion.

The two general issues are by no means sharply distinct and completely independent of each other. Nonetheless, it appears useful to consider each one of them in turn. On the one hand, there is what will be called the 'horizontal issue'. It has to do with the way in which smaller parts constitute larger entities, and the properties exhibited by the latter are related to those of the former. Among other things, dealing with the horizontal issue means that one has to discuss the well-known thesis of Humean Supervenience, according to which reality is entirely constituted by whatever occupies specific spatio-temporal points – the only additional elements being external spatio-temporal relations. This entails, in turn, that we will once again have to look at quantum mechanics. In particular, in the context of a naturalistically-inclined examination of parts and wholes and of the thesis that everything supervenes on point-like entities, the evidence already mentioned in Chapters 2 and 3, related to entangled systems and their being in some sense 'more than the sum of their constituent parts', becomes obviously important.[1] Possible conceptions of the properties of such physical systems will be discussed. Also, we will look at the relationship between physical composition in its various guises and the philosophical notions of part, whole and composition, trying to establish the extent to which the evidence coming from science determines a selection among the possibilities that metaphysicians have individuated and discussed (mostly) on purely *a priori* grounds so far.

On the other hand, there is what will be called the 'vertical issue'. It has to do with ontological priority and dependence and the questions of whether, and in what sense, the whole is prior to its parts or vice versa; it also involves an examination of the very idea of something qualifying as fundamental, that is, as such that it does not depend on anything else for its existence, identity and essential qualities. The notion of ground, as we have seen in Chapter 2, has become central in recent metaphysical analysis, and it is related to the issues concerning what is more fundamental than what, and what depends on what. Now, the once traditional monistic (originally Parmenidean) idea that the whole is, in some clear sense, more important than its parts has been recently revived and vigorously defended (by Jonathan Schaffer, in particular). And in the context of such a defence, an important role has been played, among others, by arguments based on contemporary physics and what it tells us about the mereological and dependence/priority structure of reality. An examination of these arguments is, therefore, clearly in order in the present context.

The next two sections, then, will be devoted to a discussion and critical assessment of the horizontal issue, while the following two will examine the vertical issue, also exploring its connection with the structuralist metaphysics already discussed in Chapter 3 and the supersubstantivalist ideas briefly mentioned in Chapter 4.

2.   Horizontal issue: Humean supervenience and all that

As already explained, the horizontal issue has to do, essentially, with the way in which physical theory confirms or disconfirms our intuitions and philosophically-entrenched views about composition and about the properties of composites; and this requires one to consider again the quantum mechanical facts related to the EPR/Bell scenarios examined back in Section 2 of Chapter 2.

Let us then assume that we have a clear and uncontroversial idea of what 'part' and 'whole' mean, and of how to express all mereological facts consisting of smaller parts giving rise to larger wholes.[2] One thought that appears to (more or less) directly mirror our commonsense beliefs about the structure of reality is the following: The small comes before the large, as the former constitutes the latter; and the universe is just a structured complex of smaller parts related to each other. This idea is what grounds the thesis of Humean Supervenience. In a well-known quotation from Lewis, the thesis is explained as follows:

> Humean supervenience is named in honor of the great denier of necessary connections. It is the doctrine that all there is to *the world is a vast mosaic of local matters of particular fact*, just one little thing and then another. ... We have geometry: a system of *external relations of spatiotemporal distance* between *points*. Maybe points of space–time itself, maybe point-sized bits of matter or aether or fields, maybe both. And at those points we have *local qualities*: perfectly natural intrinsic properties which need nothing bigger than a point at which to be instantiated. For short: *we have an arrangement of qualities*. And that is all. There is no difference without difference in the arrangement of qualities. *All else supervenes on that.* (Lewis [1986; ix–x, italics added])

But if Humean Supervenience were true, how could we explain the peculiar features of quantum entangled systems? Recall that it is an uncontroversial fact that measurement outcomes involving entangled particles are invariably correlated in a way that cannot be accounted for in terms of interactions between separate local quantities without contradicting special relativity. And this seems to mean that there is more to the system composed by the particles in question than what the theory tells us about those constituent particles and their properties taken in isolation – that is, that the evidence contradicts Humean Supervenience. In a bit more detail, the fact that measurement outcomes in the EPR/Bell setting are not independent means that the condition of *factorisability* fails. In the case of two particles, this condition can be expressed formally as follows:

$$P^{AB}(x, y \mid i, j, \lambda) = P^{A}(x \mid i, \lambda)P^{B}(y \mid j, \lambda)$$

where 'P' denotes probability; 'A' and 'B' refer to the two particles composing the total system; 'x' and 'y' denote the outcome at each wing of the measurement process; 'i' and 'j' denote the setup of each one of the measuring apparatuses; and 'λ' denotes the pair's state before measurement, which may encode some 'hidden variables'. As we have already seen, Jarrett [1984] argued that factorisability is the sum of *parameter independence* (independence of the outcome at one wing of the setup at the other), that is,

$$P^{A}(x \mid i, j, \lambda) = P^{A}(x \mid i, \lambda) \text{ and } P^{B}(y \mid i, j, \lambda) = P^{B}(y \mid j, \lambda)$$

and *outcome independence* (independence of the outcome at one wing of that at the other), that is,

$$P^{A}(x \mid i, j, y, \lambda) = P^{A}(x \mid i, j, \lambda) \text{ and } P^{B}(y \mid i, j, x, \lambda) = P^{B}(y \mid i, j, \lambda).$$

Without embarking in a discussion of the various reactions to this and the various proposals that have been made in the literature, let us look at one specific point of view, which is directly relevant for our present purposes. Teller [1986], siding with the majority that deems parameter independence more fundamental than outcome independence, proposed a peculiar explanation of the failure of outcome independence. He designated as *particularism* the view that the world is composed of individuals possessing non-relational properties and relations between which are nothing over and above their non-relational properties (something very close to, but not identical with, Humean Supervenience), and claimed that what the violations of Bell's inequalities show is that particularism fails in the quantum domain, and therefore one has to embrace what he calls *relational holism*. The latter is the view that certain properties are 'inherent' relations that are irreducible to the monadic properties of their relata and consequently make 'the whole (exhibiting the relation) more than the sum of the parts (the related individuals)'. Of course, the basic idea is that such a form of holism accounts for the mutual dependence of the relevant measurement outcomes: although they do not exchange information directly with one another, the two measurement events are connected by a real, physical relation.[3]

In particular, following, for instance, Gisin [2005; 5], on Teller's construal one can suggest that the allegedly problematic quantum correlations are not correlations between two (or more) events. Rather, in each case of correlated measurements, one only has a single event that manifests itself at two (or more) locations.[4] Indeed, one can even suggest (see Morganti [2009] and San Pedro [2011]) that the properties being conjectured to exist may be exploited to explain the relevant evidence in terms of a so-called 'common cause'. Indeed, a common cause is normally intended as some kind of entity determining *before measurement* the way in which the separate particles correlate when measured.[5] And since the pair's state before measurement (including hypothetical hidden variables) has been taken into account and yet the total state appears non-factorisable, the existence of such a common cause is normally ruled out in the present case. However, if what *is measured* is a holistic relation determining the particles' correlated properties, one has certainly identified something that qualifies as a cause in the required sense![6] At any rate, rather than seeing whether or not Teller's proposal (which, incidentally, is in perfect agreement with the account of quantum statistics propounded in Chapter 3) satisfactorily addresses the problem represented by the seemingly non-local character of certain quantum correlations, here we have to look at the ontological details of the picture being proposed. What does Teller's view amount to at the metaphysical level, exactly?

One first question is whether one has to agree with Teller that what is 'additional' are inherent *relations*. Another is whether one has to do with *categorical* or *dispositional* properties, i.e., with 'fully realised' properties or with 'potentialities' that are essentially related to the (possible) future occurrence of the conditions required for their manifestation.[7] As a matter of fact, Teller remains quite vague in his papers, and thus there is some space for philosophical manoeuvre here. Suppose Teller in fact intended to suggest that entangled systems exhibit categorical, irreducible relations. This is basically what the supporters of weak discernibility claim, as illustrated in Chapter 3. There, though, we provided arguments to resist this conclusion: besides being metaphysically 'unconventional', fundamental relations not dependent on their relata for their qualitative content are neither required for individuation nor, it could be contended, unambiguously identifiable as genuine properties on the basis of physical theory. If only for consistency's sake, then, let us now contend that the non-supervenient properties inherent in the relevant wholes truly are *monadic*. In a nutshell, the reasoning is, that entangled systems exhibit properties, intrinsic to the whole, that:

a) Cannot be reduced to the properties of the component parts;
b) Convey information about those parts;
c) *Need not*, because of the truth of a) and b), be conceived of as genuine relations holding between such parts;
d) Based on c) and conservativeness, *are in fact best conceived of* as non-relational.

Notice that, on the one hand, this means that particularism as defined by Teller might, after all, be preserved in spite of the evidence coming from quantum mechanics: admitting of holistic properties that 'emerge' as complexity increases, it turns out, is compatible with there being no properties that are not reducible to non-relational properties. This form of holism, it is perhaps worth emphasising, preserves the just-mentioned compatibility with the account of particle identity and quantum statistics provided in Chapter 3 (which, recall, relied exclusively on the holistic and non-reducible nature of the relevant state-dependent properties, not on their being relational or non-relational in nature).

As for the categorical/dispositional issue, consider the following question: if all we have is a categorical, irreducible property conveying at a time t_1 information such as '...has opposite spin in the x direction with respect to...', how are the correlated, local measurement outcomes at a later time t_2 determined? Recall that these outcomes correspond exactly to the monadic properties of the relata whose reality is being denied. But if the holistic correlation is categorical, it fully exists at a given time and does not in itself say anything about properties of other things at later times. At points, Teller seems realise that this constitutes a difficulty, and to suggest that the inherent relations are *dispositional* – he speaks of a 'partially effective disposition' or 'correlation-propensity' that is an objective property of the pair of objects [1989; 221–222]. To be sure, this gives us what we need: by acting 'on one of the particles', in the course of the relevant measurement procedures one affects the inherent, holistic disposition of the whole, so that the disposition 'collapses onto' separate categorical properties of the parts – that is just what its manifestation under the appropriate conditions consists of. This also provides a ground for regarding the holistic properties in question as genuine common causes in the sense suggested above – for (the occurrence of the conditions for the manifestation of) a disposition can surely be regarded as the cause of that disposition's manifestation. Should we then opt for monadic, holistic dispositions?

The response is far from being necessarily affirmative. On the one hand, the complaint that invoking holistic dispositions is not explanatory – it just gives a name to the phenomena to be accounted for – is not conclusive. On a general note, defining and characterising precise ontological categories that play certain roles in (putative) metaphysical explanations appears to qualify as more than just 'labelling' in arbitrary ways the phenomena that are to be explained. Moreover, one's mocking reaction to the doctors' explanation of the sleep-inducing properties of opium on the basis of its 'virtus dormitiva' in Molière's 1673 play *Le Malade Imaginaire* is certainly reasonable. But it is not obvious that the same goes in all apparently analogous cases. That is to say, it is less obvious than it may seem that dispositions can never be explanatorily relevant.[8] On the other hand, it is also true that postulating fundamental dispositions is not necessary in order to obtain the required metaphysical explanation. For, on a 'light' conception of dispositions, the categorical and the dispositional are identical, and the latter just coincides with a functional 'aspect' of the former.[9] One way of articulating such an identity view is by equating the manifestation of dispositions with the evolution of categorical properties according to the laws of nature. More specifically, instead of postulating basic dispositions from which the lawful behaviour of physical systems emerges, one might claim instead that things have categorical properties that, together with the laws of nature, give rise to said behaviour – so that dispositions are merely useful descriptive tools. In our case, a system, say, of two fermions in the singlet state might thus be said to simply have a categorical property – total spin 0. Given the conservation of angular momentum and the laws governing the dynamical evolution of quantum systems, this could be said to *entail* that the two fermions will exhibit different spin values upon measurement. The key contrast here is between *dispositional essentialism* (Shoemaker [1980], Bird [2005], Mumford [2004]), according to which the essence of a property is wholly constituted by the nomic or causal roles it plays and, consequently, laws of nature are just descriptions of dispositional essences of things; and what one may call *categoricalism* (see Armstrong [1997; 79]), according to which properties, or at least the fundamental sparse properties (a qualification that is especially needed, incidentally, if Humean Supervenience is assumed) cannot be dispositional, and laws of nature cannot be obtained from properties as an ontological free lunch. In favour of dispositional essentialism, it might be argued that an ontology of properties only is preferable to one of properties and laws, both for reasons of economy and because it provides a reductive analysis of laws themselves.[10] The downside is, of course, represented by the problem with the explanatory power of dispositions mentioned a moment ago. In addition, it could be maintained (following, for instance, Fine [2002]) that some natural necessities, e.g., conservation laws, cannot be accounted for in the terms of dispositional essentialism.[11] Given the foregoing, it is not clear which option should be favoured. But let us set this aside[12] as an open question concerning the properties of entangled quantum systems and get back to our main focus, the nature of part–whole relations and Humean Supervenience.

One important thing to notice is that, upon scrutiny, it looks as though the sort of holism and non-reducibility that we agreed is exhibited by quantum entangled systems is not obviously best described through the slogan 'The whole is more than the sum of its parts'. For, what emerges from quantum mechanics is that certain systems of two or more particles have properties which cannot be reduced to the properties of their components. But, the very description of the systems in questions as 'systems of n particles', seems to indicate that facts of parthood are very well-defined in the cases at hand. Why, then, should these facts be invoked to explain the above problematic evidence about properties? The problem seems to lie in the ambiguity of the slogan reported above, in turn determined by an insufficiently sharp characterisation of the intended senses of whole and part. In terms of *properties*, it is out of question that quantum mechanics renders the slogan true. Yet, in terms of *property-bearers*, it is not clearly so. These distinctions must be borne in mind, and further refined, if the philosophical import of contemporary physics with respect to our conceiving of part–whole relations and providing a formal treatment of them is to be identified with precision.

Let us then introduce some useful definitions and differentiations.[13] First of all, let us express the idea that the properties of the whole co-vary with those of the parts (i.e., supervene on them), and this points to the fact that the former are completely determined by the latter, via the following principle of compositional uniqueness for properties (CUP):[14]

> Let S be a composite system such that its component parts $x_1, \dots, x_n$ are pairwise disjoint and let $p_1, \dots, p_n$ be the sets of monadic properties of $x_1, \dots, x_n$. Then, for every property q of S, q is exemplified by S if and only if (the relevant subset of) $p_1, \dots, p_n$ are exemplified by $x_1, \dots, x_n$.

Notice that the bi-conditional in CUP is sufficient for claiming that the determination relation holds at the type-level, not only at the token-level.

(Indeed, that a whole exemplifies a certain property if and only if its parts exemplify certain other properties is trivially true when applied to particular, actual physical composites. Any such composite cannot but have certain properties of the whole (those it actually has) if and only its parts also have certain properties (those they actually have)). Now, CUP is definitely violated in quantum mechanics. For, taking for instance S_1 and S_2 to be a product state of two identical bosons and an entangled state of two bosons exactly similar to the first two, respectively, one has an immediate counterexample. The component parts of S_1 and S_2 share the same 'property-profile' $p_1, ..., p_n$, but the latter fails to determine the properties of the whole in the required way, as it determines a holistic correlation property in one case (S_2) but not the other (S_1). Thus, the above bi-conditional is not satisfied.

But let us now consider property-bearing physical systems rather than properties. The question now becomes whether quantum mechanics violates standard mereological assumptions about such systems and the way they act as parts of other, larger systems. In particular, it may be thought that quantum mechanics violates so-called 'extensional mere-ology'. Given a reflexive, transitive and antisymmetric notion of part (i.e., assuming that everything is part of itself; that if x is part of y and y is part of z, then x is part of z; and that if two things are one part of the other, then they are the same thing), one obtains extensional mereology by adding the principle that if something fails to include something else as a part, then there is a 'remainder', something which is part of the latter and has no part in common with the former. This is known as the Strong Supplementation Principle and can be rendered formally as

$$\neg P(y, x) \rightarrow (\exists z)(P(z, y) \wedge \neg O(x, z))$$

where 'P' expresses the relation of parthood and 'O' that of overlap, that is, the sharing of parts. Indeed, in the resulting system it follows that Extensionality holds, namely, that sameness of composition is necessary and sufficient for identity. Formally put,

$$(\exists z PP(z, x) \vee \exists z PP(z, y)) \rightarrow (x = y \leftrightarrow (\forall z)(PP(z, x) \leftrightarrow PP(z, y))).[15]$$

Now, is Extensionality violated by quantum systems? Literally, the principle says that a given set of things necessarily composes one specific whole and only that whole (compare with the Identity of Indiscernibles discussed earlier in Chapter 3: by stating that two things with all the same properties are identical, it says that, given a set of properties, only one thing with those properties can exist – talk of 'two things' really meaning that two names are used to refer to the same thing in the case of indiscernibility). But this doesn't tell us anything about what should happen when parts *of the same type* enter in composition relations, where, crucially, sameness of type cannot but mean *pre-composition indiscernibility*. But it is exactly *this* specific sense of sameness applied to *types* that is relevant for the understanding and interpretation of quantum wholes: recall our earlier example of exactly similar boson pairs giving rise to qualitatively distinct wholes. It follows from this that extensional mereology is not contradicted by quantum mechanics, as extensionality does *not* entail CUP, i.e., the seemingly problematic type-level claim concerning (indiscernible) wholes constituted by (indiscernible) parts.

Thus, the argument that quantum wholes are more than the sum of their parts because quantum mechanics violates extensional mereology conflates considerations concerning a given set of things 'at a time' with considerations involving indiscernible sets of things at different times, or in different parts of the universe, or in different possible worlds. Indeed, that the same pair of bosons, say, can constitute an entangled whole but also a non-entangled whole cannot but mean:

a) That literally the same two particles give rise to qualitatively different wholes, which can only happen if the two bosons are considered at different times;
b) That indiscernible pairs constitute qualitatively different wholes in different regions of space;
c) That hypothetical duplicates of an actual pair of bosons give rise to qualitatively different wholes.

But in each one of these cases extensionality holds, for the simple fact, to repeat, that it is a claim about specific wholes constituted at specific times by specific parts.[16]

Talk of duplicates (see c) above) gives us an additional hint. Suppose that, as some philosophers, think *constitution is identity*, that is, the whole is, strictly speaking, identical, the same thing as, its parts taken together. And suppose that there are certain parts giving rise to a certain whole. In this case, duplicating the parts should give us a duplication of the whole, hence qualitative uniqueness of the composite that the parts in question can give rise to. For, if the yy compose an x and they are identical to x, by Leibniz's Law, x has all the same properties as the yy. And if we duplicate the yy, the resulting yy_d will, by definition, be indiscernible from the yy. But the yy_d will be identical to, hence indiscernible from, the whole they give rise to, call it x_d. By transitivity, x and x_d will also be indiscernible. Now, this is in direct conflict with the evidence pointed at by examples such as that concerning indiscernible bosons that we have introduced a moment ago. This goes to show that quantum mechanics is in conflict with the thesis that composition is identity. It is important to see that this is by no means also a problem for extensional mereology *per se*. Indeed, it has already been argued in the past (by McDaniel [2008]), that the mere possibility of strongly emergent properties entails that composition is not identity. Define a strongly emergent property as a perfectly natural property that can be exemplified by composite material objects, and does not locally supervene on the perfectly natural properties and relations exemplified by (some of) their parts.[17] The thesis that composition is identity appears to rule out the possibility of such properties. But if the latter are possible in the quantum domain, then it looks as though the evidence provided by quantum mechanics leads to the rejection of the thesis that composition is identity. And, indeed, the sort of properties exhibited by quantum entangled system do seem, as we have just seen, to qualify as strongly emergent in McDaniel's sense.[18]

Given the foregoing, then, the appropriate reaction to the evidence concerning quantum entangled systems is not to drop extensional mereology but to say, 'Too bad for the thesis that constitution is identity'.[19]

Let us now look at Humean Supervenience. When it comes to it, the conflict between the empirical evidence and the theoretical assumptions seems evident. The thesis of Humean Supervenience – at least in its original formulation – only includes localised properties (i.e., properties of point-sized bits of matter/space–time) and spatio-temporal external relations among the constituents of the 'mosaic world' Lewis had in mind. However, at the very least, other non-localised properties corresponding to entangled systems have to be introduced at the level of what is not reducible – perhaps in the form of additional external relations, perhaps in the form of non-point-like monadic properties (recall our earlier discussion of Teller's relational holism). Now, Lewis suggested that Humean Supervenience can "doubtless be adapted to whatever better supervenience thesis may emerge from better physics" [1994; 474]. The immediate question, then, is: could one not just add quantum non-localised properties to the supervenience basis, i.e., take the evidence into account without giving up a seemingly fundamental thesis about reality?

Darby [2012] considers this possibility, focusing on the case in which holistic quantum properties are understood, *à la* Teller, as inherent relations. He argues that it is, indeed, possible to enrich the supervenience basis as suggested, but this goes against the spirit, if not the letter, of Humean Supervenience. For, while spatio-temporal relations can be systematically reduced to binary relations, in the quantum case one has to say that everything supervenes on monadic properties plus n-adic relations (both spatio-temporal *and quantum*). This is due to the fact that, exactly in the same way in which the monadic properties of two separate entangled particles do not determine the properties of the whole (the reduced states of the parts might be the same for different states of the whole), there is no way to determine the properties of entangled systems of three or more particles solely on the basis of binary relations between pairs of particles. And this seems to be more than just the statement of a fact about relations and their n-adicity. Indeed, it seems correct to say, with Darby, that the general fact about quantum systems just illustrated points to a form of holism which departs decidedly from what would seem to be the basic intuition underpinning Humean Supervenience – namely, that the world *just is* a more or less random distribution of properties at various points in space–time, *determining* that every set of properties at a point is spatio-temporally related to any of the others, and such pairwise relationships convey all the information about reality that there is. The only ways out of this seem to be the following two:

1) To argue that spatio-temporal relations too are not (always) binary;
2) To interpret Humean Supervenience as the thesis that there only exist monadic properties (*including* properties inherent to quantum wholes) plus spatio-temporal relations.

Neither option, however, appears to be very promising. As for the first, it could be based on a radically holistic/structuralist view of space–time points and relations, whereby each one of them depends ontologically on all the others. But such a view does not clearly eliminate the gap between the spatio-temporal case and the quantum case: For, it remains a fact that whatever number of objects we have, we can always analyse their spatio-temporal relations in binary terms.[20] As for the second option, obviously enough if holistic quantum properties are understood as monadic, every reference to point-sized basic facts has to be removed from the formulation of the thesis. But that was exactly the fundamental element of the Hume–Lewis framework! Therefore, this alternative too seems to be a non-starter for the supporter of Humean Supervenience.

The upshot of all this is, then, that, unlike extensional mereology, Humean Supervenience is indeed best set aside in light of the available evidence.[21]

A different issue in these surroundings which is useful to consider at least in passing – if only because it is directly related to the considerations to be made in the second half of this chapter – has to do with physical composition and the question of *when* exactly it takes place. As mentioned at the beginning of this chapter, metaphysicians have often attempted to provide general answers to questions about composition and have often found that it is difficult to find them. It is not absurd to think that this is due, at least partly, to the insufficient attention that such philosophers have paid to the empirical input of science. Let us then see whether something more can be said about this, at least as an initial step, from a properly naturalistic perspective.

The agenda was more or less set by Van Inwagen [1990], who identified three interrelated questions: the General Composition Question – 'What is composition?' [Ib.; 39]; the Special Composition Question – 'In which cases is it true of certain objects that they compose something?'; and the Inverse Special Composition Question – 'In which cases is it true of an object that there are objects that compose it?' [Ib.; 48]. The first is a general question that aims to identify essential conditions on both the composing parts and the composed objects. Answering it would also entail obtaining answers to the other two questions. Answering the other two jointly, however, does not necessarily suffice for answering the first: for, we might be able to enumerate and systematically identify cases in which things compose a whole and cases in which objects have (or fail to have) parts, without being able to tell what composition is in general. Van Inwagen expressed scepticism about the possibility of answering the General Composition Question, while others are more optimistic (see Hawley [2006]). Be this as it may, it is doubtless that most of the philosophical literature has been devoted to the special questions (see Lewis [1991], Markosian [1998], Merricks [2001], Sider [2001], Kriegel [2008], Cameron [2012] on the Special Question and Markosian [1998a] and McDaniel [2003] for discussions of material simples, relevant for the Inverse Special Question).[22] Without entering into the details of the extant debate, it will be simply assumed here that by starting from considerations involving specific cases as they are described by scientific theories, *in primis* physics, one might hope to move up towards something like an answer to the General Composition Question. What qualifies as actual physical composition, then?

In an interesting and detailed recent study, Healey [2013] points out that while philosophers have thought of material composition mostly, if not exclusively, in spatio-temporal terms, spatio-temporal considerations are often irrelevant for many types of relationships occurring in modern physics that appear to count as cases of genuine composition. For instance, in the classical electromagnetic wave conception of light, it is possible to say that the latter is composed of certain waves of definite wavelength. These waves, however, are individuated via the mathematical procedure of Fourier decomposition, which does not correspond (at least not directly) to thing-like entities in space and time. Something similar occurs in the quantum domain, in which the notion of a mixture of states – each corresponding to a definite, pure state – invites the same departure from the naïve spatio-temporally-distributed-building-blocks view of composition. Healey also notes that different distributions of pure states can generate physically indistinguishable mixed states, so that there is no direct correspondence between composing parts and composed wholes even independently of the specific nature of composition. And this happens for bits of matter in the same way as it does for light: there are seemingly incompatible, but in fact fully equivalent, ways of analysing fundamental particles (Healey considers kaons) as composed of parts and, consequently, decaying into other things.

Moving closer to our present concerns, Healey shows that the relation of *being an entangled subsystem of* can be regarded as a proper part relation but fails to define a robust compositional hierarchy due to its being state-dependent. The more general relation of *being a quantum subsystem of*, he adds, also qualifies as a more or less standard part–whole relation. However, it fails to satisfy the mereological conditions of *complementation* (if some *x* is not a subsystem of something else, *y*, then there is a third thing *z* whose subsystems are exactly the subsystems of *x* that have no subsystems in common with *y*); and *unrestricted sum* or *unrestricted composition* (every collection of subsystems of a system composes some system which has no other subsystem disjoint from all of them). The examples used by Healey are the following. On the one hand, the hydrogen atom is constituted by a proton and an electron, but in order to understand its structure, physicists normally decompose it (i.e., the corresponding Hilbert space) into a product of the position/momentum of the electron relative to the proton and the centre-of-mass position/ momentum of the atom as a whole. Since it makes no sense to identify, say, the object constituted by all the parts of the hydrogen atom minus the proton as genuine, the practice of physicists thus contradicts complementation (there is nothing like a *z* in the sense of the above general definition). On the other hand, the electron, the proton and the centre-of-mass subsystems of a hydrogen atom do not compose anything for analogous reasons: relative position/momentum is also required. Consequently, unrestricted composition seems to be in trouble, too.

To see what this entails from the philosophical viewpoint, notice, first of all, that neither complementation nor unrestricted sum are basic mereological principles. Nor are they basic within the extensional mereology that we have discussed (and, to some extent, defended) so far. Besides being dispensable from the purely formal point of view, these principles have indeed been questioned in the past because of their consequences, e.g., because they force one to be committed to the existence as genuine objects of gerrymandered 'scattered' entities. As for unrestricted sum, in particular, it is a controversial principle, the addition of which turns extensional mereology into so-called 'general extensional mereology', and which will be discussed further later on in this chapter.

A second set of considerations must be made, though, concerning how (un)restricted composition is. In his discussion of kaons and the Standard Model, Healey seems to more or less take for granted that physical composition takes place when there are forces connecting otherwise independent systems. Does this mean that interaction governed by a fundamental force is sufficient for composition? If so, what follows from this idea? Given that gravity acts among any pair of massive particles and never reaches intensity 0 (i.e., its range is infinite), if the hypothesis being considered were true, every subset of the things (provided with mass) that exist in the universe would actually constitute a whole. Notice that the restriction to entities with mass suffices for not inferring from this that unrestricted composition holds: for, it makes it simply false that for *any* set of entities there is a whole constituted by the entities in that set – the set might comprise one or more massless entities that, as such, do not 'feel' gravitation. At any rate, gravitational interaction cannot be all there is to physical composition, for instance, because it does not account for entanglement, and entangled systems certainly count as composites.[23] Let us then briefly look at the other kinds of forces and interactions posited by contemporary physics. Now, the strong force definitely determines composition: it is the force – studied by quantum chromodynamics – that governs the way in which quarks make up all the fundamental material particles. Indeed, quarks are supposed not to exist if not into the composites they give rise to! But, again, this leaves out a long list of cases that might plausibly be regarded as cases of composition, including typical entangled systems. What about electromagnetism and the weak force? The latter affects all fermions, i.e., bits of matter, and essentially determines decay. Hence, besides not accounting for composites made up of bosons, it seems to have to do with decomposition rather than composition (even though subatomic particle decay is a typical process of entanglement production, hence it does create genuine wholes). In view of this and, more importantly, of the fact that electromagnetism and the weak interaction are now believed to be two aspects of the same interaction (one that would effectively be manifested in the form of a unique force if energy were high enough), we can focus on electromagnetism only, that is, on the study of the force that acts on all charged particles. Since electromagnetism is involved in chemical bonds, such as the covalent bond whereby two atoms literally share one or more pairs of electrons, it definitely grounds at least one type of seemingly genuine composition. But other electromagnetic phenomena also appear to give rise to clear cases of composition. For instance, entangled photons can be produced via parametric scattering, that is, roughly, the splitting of bosons by crystals with specific features in such a way that (due to energy conservation) the produced photons have correlated polarisations. And this would be an electromagnetic process determining the creation of a genuine physical whole. More generally, it looks as though electromagnetism can create part–whole relations between elementary bosons, even though these are not electrically charged in all cases, and thus the force may not act directly between them. Notice that the electromagnetic (or electroweak) force has infinite range, exactly like gravity. Therefore, in this case, too, it seems that interaction occurs between any pair of particles with the appropriate characteristics, and a 'restricted unrestricted composition' appears to hold.

Without adding more details, let us draw some general conclusions from the above – admittedly sketchy – considerations. It can be plausibly suggested that, roughly, one has (ordinary – recall Healey's argument to the effect that physical composition is more than just spatio-temporal composition) physical composition when:

a) Atoms share electrons (i.e., chemical bonds are established) – which is, in turn, the ground for composition at higher levels of complexity, or
b) Particles interact by exchanging gauge bosons, or
c) Particles become entangled via some physical process.

(As we have pointed out, none of these justifies the belief in unrestricted mereological composition.) Is the above all there is to say about what sums count as genuine physical wholes? On the one hand, Healey refers quite vaguely to what 'is natural' [2013; 11] and what is regarded as a whole by scientists, without providing more detailed philosophical argument. This seems to indicate that there is further work to be done by philosophers interested in these issues – for instance, to identify and make explicit any further constraints that might have to be satisfied for interactions truly to give rise to genuine physical wholes. On the other hand, metaphysicians might insist on their *a priori* views contending, for example, that the above does nothing to show that nihilism is false and, strictly speaking, nothing truly composes anything. To this, it cannot but be replied that the underlying attitude is definitely against the spirit of the naturalistic methodology. If there is anything tangible that could be brought to bear on the nihilism issue in metaphysics, is evidence such as the above.

These issues are, no doubt, important, and there certainly are other themes of interest in these surroundings. However, for the time being, we can conclude here our illustration and discussion of the horizontal issue with the part–whole relation from the viewpoint of naturalised metaphysics, and move on to a general evaluation.

3.   Evaluation and proposal

One first, fundamental thing to notice is that certain philosophical views are more or less conclusively ruled out by their lack of fit with the evidence alone. These include what we called 'compositional uniqueness for properties'; the thesis that composition is identity; the view that the concept of composition exclusively denotes spatio-temporal relations; the idea that composition relations are objective and unique; mereological frameworks richer than extensional mereology (in particular, those including complementation); the thesis of unrestricted composition; and the form of particularism underlying Humean Supervenience. Simplicity and modesty additionally make a modified Humean Supervenience including non-point-like (possibly relational) quantum properties in the supervenience basis unappealing. On the other hand, a form of holism that acknowledges the possible existence of inherent properties of certain wholes that neither supervene nor depend upon localised properties of the parts makes sense of the available evidence and, more generally, fares well on all counts.[24] Also, conservativeness seems to invite one to make such holism depart as little as possible from traditional particularism, and so regard the inherent properties of the wholes as monadic. While whether these properties are dispositional or categorical (and, thus, what the nature of, and role played by, laws of nature exactly are) seems instead an open matter, considerations of empirical accuracy and fruitfulness recommend that the inherent properties be regarded as genuine constituents of wholes such as entangled systems. For, only then do they appear able to play certain ontological, hence explanatory, roles – most notably, as common causes for correlated outcomes in EPR/Bell scenarios. Again based on conservativeness, simplicity and generality together with compatibility with the empirical data, it also seems advisable to regard (or continue regarding) extensional mereology as a formal framework adequate for the description of part–whole relations as these are realised in the physical world. Lastly, as for the infamous composition questions, physics appears to give at least some useful indications with respect to how to tackle them. And while much more can and should be said about this, it seems in any case possible to suggest that composition really consists in interaction via a fundamental force and, more generally, in the 'sharing' of something: gauge bosons in the case of elementary particles, particles in the case of chemical bonds (which, as mentioned, can also be regarded as the ground for cases of physical compositions at higher levels of complexity), and, so to put it, of wavefunctions in the case of entanglement.

Having said this, we can now move on to what we called the 'vertical issue' with part–whole relations, which is also the last case study that will be presented in this book.

4. Vertical issue: grounding and fundamentality

As discussed in Chapter 2, the notions of ground and ontological dependence have recently acquired the centre of the stage in metaphysical discussion. Given that, as we do in the present work, we find this evolution justified and indeed welcome, it is certainly necessary to discuss such notions in all of their aspects here. Especially so when they turn out to be directly relevant for the specific issues in naturalistic metaphysics that we are interested in at the level of case studies. Indeed, this is exactly what happens when it comes to the issues that constitute the, object of this and the next section. Let us see this in more detail.

4.1 Priority monism

The *monist* view that it is the universe as a whole that is ontologically prior and independent, while its parts are derivative and dependent, was held by the likes of Parmenides, Plato, Plotinus, Proclus, Spinoza, Hegel, Lotze and Bradley. Once dominant, it was then set aside in favour of the *pluralist* idea, perhaps better supported by the key presupposition, underlying Newtonian physics, that what is truly fundamental are elementary entities more or less like Democritus' atoms. Nowadays, however, monism has re-emerged as a relevant thesis in metaphysics. In the form of *existence monism* (Horgan and Potrç [2000], [2008]), it is the thesis that there is only one whole; in the form of *priority monism* (Schaffer [2010]), it is the thesis that both the cosmos and its parts exist, but the former is prior to the latter. There is also the Kantian form of monism developed by Kriegel [2012], according to which the world decomposes into parts insofar as an ideal subject would recognise it as divided into parts, but not in a mind-independent manner.

In what follows, we will focus on Schaffer's position, taking it for granted that it is the form of monism that minimises the departure from common sense and one's commitment to controversial philosophical assumptions about reality and our knowledge of it. Also Schaffer argues in favour of priority monism on the basis of considerations related to the physical world and physical theory. It is thus clear that a discussion of his arguments is in order in the present context. Schaffer assumes that ontological priority relations form a well-founded partial ordering, that is, that they are irreflexive, asymmetric and transitive and give rise to series of dependence- (and, therefore, priority-) relations that terminate at some fundamental level of entities that do not depend on anything. This amounts to the endorsement of *foundationalism*, the claim that there must be a 'ground of being', and chains of dependence can be neither infinite nor circular. Assume next that composition is not identity (something which we already found plausible) and that the 'tiling constraint' (the idea that the basic objects together constitute the world in its entirety, and no basics are related as whole to part) holds. All this entails that monism (i.e., the claim that there is exactly one basic concrete object, the universe as a whole – formally, $\exists!xBx$, or, equivalently, $\wedge Bu$, where 'B' denotes the property of being basic and 'u' the universe as the all-comprising object) and pluralism (i.e., the claim that there are two or more basic objects – formally, $\exists xy(Bx \wedge By \wedge x \neq y)$) are mutually exclusive and exhaustive theses. Against this background, Schaffer provides (at least) six arguments in favour of priority monism. Let us illustrate each one of them in turn and then critically assess them in the same order.

1) *The argument from quantum mechanics.* Schaffer exploits the – by now well-known – holistic consequences of facts about quantum entangled systems to argue as follows. Since it is reasonable to believe that

entanglement is a pervasive and ubiquitous phenomenon, so that the whole cosmos is in a non-factorisable state (as contended, for instance, by Zeh [2004; 115]), it is also reasonable to conclude that the cosmos is the truly fundamental physical system. For, entanglement is only truly explained by giving ontological priority to entangled wholes rather than to their parts..

2) *The argument from the logical asymmetry of emergence.* Generalising from (1), Schaffer focuses on emergence, that is, on the possibility of new properties arising as complexity increases which are not determined by the occurrence of ontologically more fundamental properties, as a universal phenomenon. Pointing out what he dubs the 'asymmetry of emergence' – whereby the whole can have features that the sum of its parts does not have, but the parts cannot have features that the whole comprising them fails to have (there is no such thing as 'submergence') – Schaffer again concludes that the whole is fundamental, while its parts are dependent.

3) *The argument from the possibility of gunk.* The hypothesis has been entertained[25] that everything is divisible, and the universe is consequently made of 'atomless gunk', an infinite series of progressively smaller parts. Schaffer argues that, while the monist has no trouble accounting for such a possibility, as the infinite series of divisible parts is still contained in the unique unitary cosmos that he or she regards as the fundamental object, the pluralist simply cannot do so. For, if parts are prior to wholes, gunk entails that the assumption of pluralism violates foundationalism, as instead of an ultimate, well-defined set of independent entities, the pluralist finds an infinite series of dependent ones. In connection to this, Schaffer additionally proposes an 'asymmetry of existence', according to which atomless gunk is possible, but 'worldless junk' (everything is a proper part of something) is not, as there always exists a 'totality' object that comprises everything.

4) *The argument from the internal relatedness of all things.* Schaffer [2010a] also offers the following piece of reasoning: contrary to the Humean principle of free recombination, which one would expect to hold in a pluralist world populated by a multitude of basic, hence independent, entities, all things are internally related in ways that make them mutually interdependent; but such ubiquitous interdependence is best explained by endorsing priority monism. Schaffer presents three plausible candidates for pervasive internal relations, i.e., relations essential to their relata and such that they would equally hold between any duplicates of such relata. The first is causal

connectedness in the framework of dispositional essentialism (a thesis we have already encountered): everything is causally connected in the causal structure determined by the cosmos in the course of its history, but causality just is the manifestation of essential dispositions, hence everything is essentially connected. The second is the world-mate relation in counterpart theory: according to counterpart theory, the identity of each entity in each possible world is unique and fully determined by the fact that the entity in question exists in *that* specific world; but this means that only the set of everything that exists in a world, which is what makes the latter the world it is, can individuate any entity inhabiting that world; and this, in turn, means that each entity in each world is related to all other entities in that world. The third is spatio-temporal distance in the context of structuralist supersubstantivalism: if concrete objects are identical to regions of space–time (the supersubstantivalism we have already looked at earlier), and distance relations are essential to regions of space–time (holistic structuralism with respect to space–time), then it follows that all things are essentially spatio-temporally related in the way they are.

5) *The argument from space–time as the fundamental substance.* Relatedly, Schaffer [2009] argues that 'space-time is the one substance' based on considerations of ontological economy and explanatory power as well as on the plausibility of a holistic, field-theoretic understanding of quantum field theory and General Relativity. (Compare again with the discussion of supersubstantivalism in Chapter 4, Section 3.)

6) *The argument from the action of the whole.* Lastly, Schaffer [forthcoming] assumes the Leibnizian view of substances (i.e., non-dependent, integrated wholes) as what evolves according to fundamental laws and the Russellian view that only the cosmos evolves in this way, and again concludes that the whole universe is the only fundamental thing. The former view is supported by the seemingly forceful intuition that fundamental laws must directly apply only to what counts as a fundamental entity. The latter, by the observation that truly isolated systems are rare if not impossible, and thus the behaviour of something less than the whole universe never really corresponds to our predictions based on fundamental laws. A weaker version of this argument has it that – even though other things might do so – the universe surely counts as a substance because it follows the laws of nature, and since no part of a substance is itself a substance, this

suffices for concluding that the universe is the only fundamental entity.[26]

What follows from the above considerations? Should we endorse monism? First of all, notice that the tiling constraint and the view that composition is not identity might be questioned. However, first, it is very plausible that basic entities cannot be parts of equally basic entities (for then it would be natural to identify the part and its complement, rather than the part and the whole, as basic entities); and it is equally sensible to think that, given what is basic, what is derivative is somehow also given – at least in the mereological sense. Therefore, there is no reason for dropping the tiling constraint. As for composition as identity, we have already provided reasons for rejecting that hypothesis. Here, we will just add that the assumption that composition is not identity is required in order to have a well-defined opposition between distinct views to begin with, and should thus be granted at least for the sake of argument. But, let us take a critical look at the above arguments in favour of priority monism.

Starting from Schaffer's considerations concerning quantum physics, the idea that the cosmos as a whole is entangled might plausibly be rejected.[27] But let us grant that assumption. Schaffer himself suggests that the pluralist has the option of including, in the way discussed earlier, holistic quantum properties in the set of the basic entities, although this entails that not all basic entities are at the same ontological level. (Some emerge at higher levels of complexity than others.) He objects to this based on two considerations: first, relativistic quantum field theory makes it possible that particles will not be retained in future physical theorising, which also affects the ontological status of allegedly fundamental, but irreducible, relations among them; secondly, if one regards entanglement relations as fundamental, ontological unity cannot be preserved because what would *prima facie* seem to be identical relations (of the total system) in fact coincide with a number of different relations, roughly dependent on the number of constituents of entangled systems. However, the first of these two points seems to rely on a confusion between considerations and levels that had better be kept apart. While it is true that the particle concept is difficult to retain in the field-theoretic context, it is also unclear – as we have seen in Chapter 3 – what the ontology that such a context suggests is. At the same time, there are good reasons for claiming that entanglement relations remain ontologically basic when one switches to the relativistic quantum field-theoretical perspective. As for the second point, it appears by no means clear that providing a unitary account of entangled relations is a problem for pluralists any more than it is for monists. To begin with, the nature of entanglement is only easily understood in the two-particle case, and it is an open experimental question how to even recognise when many-particle systems involve entanglement among all their components, or just a subset of them. Simply mentioning the total property of the system unduly hides this important complexity. In view of the latter, the search for physical unity may even turn out to be unmotivated. But suppose such search is instead motivated. Besides the fact that a sufficiently unitary analysis of entanglement relations, based on a few types of basic entanglement relations, might turn out to be possible upon further scrutiny, simply inverting the direction of ontological dependence does not, by itself, make the nature of entanglement relations any easier to identify and account for. One may retort that this might be so, but postulating the existence of holistic and monadic entanglement properties (rather than relations) along the lines that have been suggested earlier in this chapter leads instead directly into monism. But this is not true. As a matter of fact, one can legitimately continue to maintain that, reality is best described in pluralist terms – although not in those of the particularist version of pluralism presupposed by Humean Supervenience.

Moving on to the supposed logical asymmetry between emergence and its converse, granted the possibility and actuality of genuine emergence it seems fair to claim that from such asymmetry it does not follow that the whole is prior to the parts. Why should the fact that the whole is richer than its parts in the sense pointed out by Schaffer entail that the direction of dependence is not from parts to whole but the converse? In fact, it looks as though the whole and its properties are, in any case, dependent on the parts and their properties for their existence (and, therefore, for their identity: see Lowe [2010]), as the former might fail to exist, while the latter exist, but not vice versa. That is, emergent properties are additional 'aspects' of the whole that are created *once* certain initial parts, properties and external relations are given (together with the appropriate form of interaction). After all, as it should appear obvious by now, the pluralist thesis does not in any way require that all facts about complex wholes be already contained in facts about their parts.

Let us now look at Schaffer's third argument, based on the possibility of gunk. In his defence of monism, as we have seen, Schaffer considers the conceivability – and also scientific respectability – of gunk, as opposed to the implausibility of junk, as a key element in favour of the monist thesis. However, junk is as conceivable as gunk. To begin with, just imagine an infinite series of objects each one contained in a larger one, with no 'top' object containing all the smaller ones.[28] For instance, it is a scientifically respectable hypothesis that we inhabit a universe such that

a) Space is infinite in size and almost uniformly filled with matter;
b) There is, in fact, an infinity of parallel universes, and
c) There is also an infinite series of multiverses, each one containing a universe like ours and universes parallel to it.[29]

If reality could be such a series of mereological layers going on *ad infinitum*, is it not plausible to take this to entail, or at least, warmly recommend, the abandonment of the mereological axiom of unrestricted composition, which is *the sole responsible* for the belief in the necessary existence of a universal object?[30] Especially in view of what we said against such axiom in earlier parts of this chapter, the answer to this question should appear obvious.[31] At the very least, it seems that one is left at an impasse whereby pluralism might appear in trouble given certain conceptual and empirical considerations about what is possible, but so does monism. (This may lead one to give up foundationalism itself, but we will not pursue this line of thought here.)[32]

Let us consider next Schaffer's argument from the internal relatedness of all things. Schaffer's move from failure of free recombination to monism is based on the idea that basic entities cannot be modally constrained. But the latter is not an obvious truth. Indeed, the pluralist can restrict the independence of basic entities so that it does not include free recombination. For instance, he or she could say that the basic entities do not depend on any other entity for their existence and identity, but are nevertheless constrained by the laws governing reality when it comes to getting together and composing wholes. Regardless of whether or not one is convinced by this, the pluralist can, in any case, focus on the other part of Schaffer's argument, providing positive support for the thesis of universal mutual internal relatedness. With respect to it, it is sufficient to notice, in a completely general way, that all three philosophical views that Schaffer makes reference to (dispositional essentialism, counterpart theory and structuralist supersubstantivalism) can be rejected and are, in fact, object of intense philosophical dispute. Moreover, with respect to the argument from dispositional essentialism, it is far from obvious that the diachronic fact that every existent concrete entity is connected to every other because they all derive from the Big Bang is sufficient for a synchronic claim of dependence on the whole.

The argument from space–time as the fundamental substance just needs to be mentioned here, as the pros and cons of supersubstantivalism have already been illustrated in the previous chapter, and the addition of structuralist geometrical essentialism does nothing but further detract from the (initially not particularly high) plausibility of the view.

With respect to the argument from the action of the whole, lastly, Schaffer himself acknowledges that it might just be a contingent fact with no metaphysical import that there are virtually no isolated physical subsystems of the universe which are correctly described by fundamental laws. Hence, the Russellian assumption that only the cosmos counts as a substance might be questioned, for surely it cannot be intended as a merely contingent claim if the overall argument is to work. To the weaker version of the argument that, as we have shown, Schaffer offers in reaction to this, it can additionally be objected that, once it is granted that things other than the universe may evolve in agreement with the fundamental laws, the idea that the universe is fundamental loses its justification. Slightly differently put, in the weakened Russellian scenario, the universe may well count as a substance in the sense that it qualifies as an integrated whole with 'lawful behaviour', but not in the sense that it is a fundamental thing that, as such, cannot contain other fundamental entities. Of course, this amounts to saying that the traditional views on basicness and the direction of ontological dependence and priority should be preserved.

Having concluded the critical examination of the peculiar view of part–whole relations constituted by priority monism, in this case, too, we can now move on to a more general overview and evaluation of what we called the 'vertical issue'. Before that, however, it is useful to consider an interesting analogy that has not been explored in detail so far in the literature: namely, that between priority monism and the metaphysical form of structuralism already encountered in Chapter 3.[33]

4.2 Monism and structuralism

As we have seen, ontic structural realism adds to the epistemic structural realist claim that, in spite of theory-change in the history of science, we can be realist about something (that is, the concrete counterpart of preserved theoretical structure), the idea that structure is all there is. We have also seen that this is allegedly motivated by indications coming from contemporary science, mainly physics, and is thus taken to warrant a sort of consilience in favour of a structuralist metaphysics. Independently of the criticisms of the view and of the arguments in its favour formulated in earlier sections, it now becomes interesting to look in a bit more detail at the structuralist metaphysics – especially so because, in at least some variants, it turns out to be surprisingly close to the sort of monism we just discussed. Even though monism itself did not appear particularly compelling, such an analogy might motivate possibly fruitful further analyses of the two positions and of their mutual interrelations, and appears, in any case, of interest in the present context.

To begin with, let us make it explicit that metaphysical structuralism is, as a matter of fact, a family of philosophical views, sharing the assumption that relational structure is (epistemically and) metaphysically prior, rather than a unique, well-defined view. One possible classification is as follows ('MS' standing for 'metaphysical structuralism' from now on):

a) Strong MS: the radical, eliminative view according to which there are only relations and objects can be dispensed with as emergent or even epiphenomenal;
b) Moderate MS: objects and relations are on a par, as all properties are, at root, relational in nature, but objects also exist as 'place-holders' in the relevant structures;
c) Mild MS: objects, properties and relations are on a par, like in the moderate version but with the modification that some properties might be monadic; identity facts, however, are always contextual.

Now, we have already considered possible sources of perplexity with respect to eliminativism about objects and all their monadic properties and contextualism about all identity facts – that is, with respect to strong and mild MS. But moderate MS does not seem to be a particularly appealing option either. The position was originally proposed (by Esfeld and Lam [2006]) with a view to accounting for the pervasiveness of relations allegedly pointed out by contemporary physics while avoiding the problem raised by the 'relations without relata' objection. However, notice that, first, the reduction of all properties to relations is still presupposed and still requires a compelling argument. Moreover, two other potential difficulties emerge. One is that, since they are mere placeholders, objects become analogous to Lockean bare particulars – which, of course, is in no obvious way better than the postulation of primitive, non-contextual identities. It is not an option to claim that, unlike in the case of bare particular ontologies, the identity of objects is derivative because extrinsic and contextually determined. For, if objects as placeholders are 'mere matter' even more deprived of characterisation than bare particulars due to their lack of identity, what is it exactly that is being postulated? And how is the resulting view any better than a more straightforward commitment to relational structure only? The second difficulty has to do with ontic dependence: if, as moderate metaphysical structuralists seem to think, objects and relations are ontologically on a par, neither element being prior to the other, then objects depend on structure for their existence and identity, but structures depend on the objects that act as placeholders and 'realisators' for them. But this mutual ontic dependence appears suspicious. Fine [1994] and Lowe [2003], for instance, claim that ontic dependence must be well-founded because it is essentially the metaphysical correlate of a relation explaining the identity of things at the epistemic level.[34] Obviously, supporters of MS can contend that mutual ontic dependence may be accepted at least in some cases, and can be framed in an adequate conceptual framework. The worry with this is that it is not clear that moderate MS is convincing enough in the first place to justify this move. In particular, it is not clear that the metaphysical scenario proposed by moderate structuralists is better than what they want to avoid, namely, a scenario where relations simply do not have objects as relata. As a matter of fact, that relations need relata is far from being agreed upon by all opponents of MS, let alone by supporters of the view!

This is the important point for present purposes. While one obvious option to choose as a reaction to the above is to insist on strong MS intended as the view that there are self-subsistent relational structures not dependent on anything like objects and their intrinsic properties, there are two other options. First, one could claim that, indeed, relations must (and in fact do) have relata, but these relata can always be analysed in terms of further relational structure, and there is no need to expect the analysis to terminate. Ladyman and Ross, for instance, claim that "the relata of a given relation always turn out to be relational structures themselves on further analysis" [2007; 155], so accepting the view that – in slogan form – 'it's relations all the way down'[Ib.; 152].[35] Now, while this hardly solves the problem of defining an appropriate metaphysics for MS if one presupposes pluralist foundationalism, things stand differently once one drops that assumption. In that case, it might actually be true that it is relations all the way down, and yet at each level of analysis specific relata can be found for the relevant relations. Also, the structuralist is not required to locate the fundamental anywhere: rejecting foundationalism, he or she can allow for an infinite hierarchy of structures in the upward direction as well. Suppose the structuralist decides to stick to foundationalism and claim that, while relational structure is always downward-analysable in terms of further relations, the truly fundamental object is the world-structure, i.e., the set of all the relations that hold between some relata. This is, of course, exactly what priority monism claims, and so it turns out that, if one presupposes metaphysical foundationalism, metaphysical structuralism might be regarded as a form of priority monism.

In light of the above, were compelling independent arguments for each position available, MS and priority monism might reinforce each other, the former providing a science-based argument for the latter, and the latter offering a precise ontological setting for the former. Here, however, it cannot but be stressed that, as things stand, the two views taken on their own do not seem particularly compelling. In particular, MS seems to be an even more controversial thesis than priority monism – at least in its strong version, which seems to be presupposed in the relations-all-the-way-down perspective: for, unlike priority monism, strong MS in the form currently being considered admits of no objects and no intrinsic properties at any level of ontological analysis (if not in the form of non-fundamental entities or even just mere epiphenomena).[36]

5. Evaluation and proposal

In terms of fit with the data, generality and fruitfulness, priority monism is, no doubt, an appealing view. It is clearly argued for with an eye to finding support for it in science, and it provides an elegant account of reality at various levels. However, it is less clear that it fares well with respect to consistency. For, recall what was said with respect to the fundamentalist assumption and gunk versus junk. Once it is accepted that nothing should lead us to insist on the pluralist presupposition that there is a fundamental level of small things, it seems unwarranted to just resort to our entrenched mereological assumptions with a view to insisting that there certainly is a universal object that counts as fundamental. Something similar holds for modesty, especially insofar as it involves the dependence on plausible assumptions: we have contended that each one of the arguments Schaffer presents in support of his views can be rejected, and that most of them are far from compelling from the philosophical and/or scientific viewpoint. If this is so, why then throw away our pluralist beliefs so radically, instead of just accepting some form of holism but not the idea that the direction of ontic dependence in fact goes in the opposite direction to what we thought so far? With this, of course, we have arrived at the criterion of conservativeness. Pluralism seems to fare better on this count. We have illustrated, though, that pluralism can be preserved but has to be modified at least to some extent (so that it goes along with an appropriate form of holism). In view of this, further study of science in parallel to certain metaphysical options is certainly required in the present case. It is in this sense that, although our conclusions were sceptical, the parallel between metaphysical structuralism as an allegedly plausible view on reality as described by our best current science and priority monism/anti-foundationalism was regarded as worth drawing and exploring.

6. Conclusions

Overall, our claims in this chapter can be summarised as follows.

First, it is certainly true that the evidence related to quantum entangled systems demands ontological revision and the commitment to some form of holism. But while it forces us to essentially abandon Humean Supervenience and superficial forms of particularism, such evidence does *not*:

a) Entail a metaphysics of relations only, nor a commitment to irreducible quantum external relations (the holistic properties we have to recognise might perfectly be monadic properties);

b) Entail that mereology and its formal apparatus is no longer useful and appropriate for accounting for part–whole relations – these might be different from what they were in the classical domain, in that they are neither objective and univocal, nor uniquely spatio-temporal; and might also fail to include certain allegedly basic axioms – but all this, and also the idea of properties and parts that emerge as complexity increases – can perfectly be accounted for in mereological terms.

Second, we individuated a number of options, but left the choice open as to the precise ontological status of quantum holistic properties, especially with respect to whether or not they should be regarded as dispositions. In view of the need for explanation of the EPR/Bell scenarios, however, we recommended a realist stance towards said properties, to be conceived as genuine constituents of physical systems[37] and, therefore, possibly as common causes for the peculiar correlations exhibited by entangled systems.

Third, we argued against priority monism – the view that both the whole universe and its parts exist, but only the former is fundamental – and against metaphysical structuralism conceived of as one of its possible forms (and a pretty radical one, indeed). It was contended that the various elements put forward in support of these views are not conclusive, and thus conservativeness recommends one to remain a pluralist about the fundamental. However, at the same time, while we did not find any positive reasons for being 'metaphysical infinitists', and so deny the existence of a 'ground' level as well as that of a 'top' level, we recognised that not taking foundationalism for granted opens up an important space of conceptual resources and might consequently turn out to be a philosophically fruitful choice in the future.

With this, our third case study, and with it the entire book, comes to a close.

Notes

1. After all, Einstein's locality assumption is analogous in spirit, if not in letter, to the basic intuition underlying the thesis of Humean Supervenience.
2. There are several questions that emerge already at this level, namely, when the abstract formal schemes of mereology are applied to metaphysics. For instance, is there an additional composite object for any given group of objects (Putnam's question again!)? Is the composite whole identical to its constituent parts? Is there any composite object at all? We will not attempt here to provide general answers to these questions and related ones. Instead, we will limit ourselves to making less systematic – but, hopefully, more amenable to real discussion – remarks and considerations as they will emerge from a parallel examination of the evidence coming from physics and the conceptual schemes of metaphysics.
3. For discussions of holism and non-separability in quantum mechanics, see Healey [1991] and [2009].
4. Compare also Lange: "the weirdness pertains to the character of the events themselves rather than to their causal relations, spatiotemporal locality is satisfied" [2002; 294]. It thus seems that in quantum mechanics certain 'things' exist that are 'basic units' regardless of the fact that they are 'extended' in regions larger than single points in space, so that a sort of unity and fundamentality exists that is different from the one we usually assume, which takes space–time points as minimal elements. Howard's words apply here: "The

mistake is in thinking that the structure of the space-time manifold can be insulated from the nonseparability that affects the rest of our physics, so that this manifold stands alone as a ground of individuation" [1989; 249]. What about the process seemingly being instantaneous or, at least, faster than light, as it takes place at two locations at the same time? Chang and Cartwright [1993] address this worry. Considering a measurement of position, and emphasising that any exact outcome is perfectly anti-correlated with a negative result for localisation at every other point in which the system could have been detected given how 'spread out' the initial wave-function was, they conclude that what we are requiring an explanation for is, in the end, nothing but the dynamics of the collapse of the wavefunction. The observation that the notion of collapse is fundamental in standard quantum mechanics (and variants of it sharing the idea of discontinuous evolution) leads them "to reject the finite-speed propagation requirement for these special kinds of quantum measurement processes" [Ib.; 183].

5. The principle of the common cause, according to which – roughly – a correlation between events A and B indicates either that A causes B, or that B causes A, or that A and B have a common cause, dates back to Reichenbach [1956] and it has been object of intense discussion. For an overview and a list of references related to quantum mechanics as well as to other domains, see Arntzenius [2010].

6. It must be stressed that these 'post-measurement' common causes violate 'causal separability' – the requirement that an event A can be the cause of another event B only if A has a part entirely in the past light-cone of B which entirely causes B. Still, it looks as though the failure of causal separability is preferable to the violation of locality, if only because non-separability is a fundamental fact about the physical systems under study anyway.

7. These definitions are very rough. There is a huge debate concerning the definition of a categorical as opposed to a dispositional one, how to analyse dispositional properties, and the relation between the categorical and the dispositional (Identity? Reducibility of one to the other? Coexistence?). We will by and large avoid addressing these issues in what follows, as our discussion will not be affected by one's specific views on these matters. It remains uncontroversial, nonetheless, that the concept of a dispositional property plays (or, at least, may play) an important role in the interpretation of quantum mechanics, also independently of the specific features of entanglement. See, for instance, the propensity interpretations of the properties denoted by quantum probabilities proposed by Popper [1957], Maxwell [1988] and, more recently, Suarez [2007].

8. For an overview of options and a defence of dispositions in this sense – i.e., as not necessarily to be reduced to their 'categorical bases' – see McKitrick [2005].

9. There are various forms of identity views. Armstrong, Martin and Place [1996] and Mackie ([1973], [1977]) defend a 'type-identity theory' according to which each kind of disposition is identical with its causal basis. Mumford [1998] endorses instead a 'token-identity theory' according to which any instance of a disposition is identical with a specific causal basis. A 'functionalist view', according to which a disposition is a second-order property of having some causal basis or other, is also available (Prior, Pargetter and Jackson [1982]).

10. I say 'might' because the issue is in fact open to debate. Mumford [2004], for instance, explicitly contends that dispositional essentialism obviates the need for laws; but Bird [2007] disagrees.

11. For the view that reality is basically dispositional, see Blackburn [1990] and Holton [1999]. For the opposite view that everything that is real is *ipso facto* categorical, see Strawson [2008].

12. There are many other things to say about dispositions, for which there is no space here. For example, can dispositions be bare (i.e., lack a categorical causal basis)? What is the relationship between dispositions and their bases? How are ascriptions of dispositional properties to be analysed? For an overview and references, see Choi and Fara [2012].

13. The following discussion of quantum ontology and mereology is indebted to Calosi, Fano and Tarozzi [2011].

14. Calosi, Fano and Tarozzi call it 'Property Compositional Determinateness' [Ib.; 1746].

15. This uses the notion of proper parthood, denoted by 'PP', which corresponds to something being part of something else but not vice versa. Explicit reference to complexes is required in order to avoid triviality: if two simples are identical they clearly share the same parts, i.e., the unique simple identical to each.

16. Notice that the same reasoning applies if it is assumed that the bundle theory is correct, and that objects really are just mereological composites of properties (see Paul [2002], [forthcoming]). While it is true that this theoretical framework is such that sameness of parts is *ipso facto* sameness of properties, it remains the case that, given the composing properties, the qualitative uniqueness of the composed wholes does *not* follow: it is still compatible with extensional mereology that *in some cases* the composed wholes exemplify properties additional to those exemplified in the initial, pre-composition state of affairs.

17. This follows McDaniel [2008] with some minor differences. First and foremost, it does not assume that a fundamental level of basic parts exists – more on this later. As for perfectly natural properties, it suffices here to take them to be genuine properties of physical objects 'certified' by current science.

18. One may think that the properties inherent in entangled wholes are perfectly natural external relations, and so the duplication process should include them, and thus composition can be identity after all. But this would be wrong. For, remember, the component parts are not in an entangled relation before some natural process makes them compose a certain whole in a particular way; and it is such pre-composition parts that get duplicated in the thought experiment supposed to show that certain common assumptions about the part–whole relation fail in the quantum domain.

19. For, an attempt to defend the thesis that composition as identity does not necessarily rule out strongly emergent properties, see Sider [forthcoming].

20. For a perspective of this sort, developed with a view to accounting for cases of seemingly absolute indiscernibility in the terms of the bundle theory of object constitution, see Demirli [2010].

21. Unless one follows Loewer [2004] in preserving Humean Supervenience as a claim about points in configuration space rather than physical space(-time). This proposal is interesting, and certainly relevant for those who have independent reasons for believing that configuration space is the truly

fundamental space. It also ties nicely with the framework proposed, for instance, by Barbour. However, we will not discuss it in detail here, but rather set it aside as a theoretical possibility.

22. There is a variety of positions in the literature. Lewis [1991], for example, thought that any arbitrary collection of particulars composes a unique particular. Others, following Van Invagen's restrictions on what counts as a genuine whole, endorsed the nihilistic view that composition never occurs (see, for instance, Rosen and Dorr [2002]). Something more on this will be said later.

23. One could also say that gravity does not compose genuine wholes because gravitons, the mediators of gravitational interaction, are still hypothetical entities, while in the case of the other forces exchanged gauge bosons have been identified as actual entities. Also, gravitons do not appear in the Standard Model of elementary particles. Another relevant fact is that general relativity reduces gravitational attraction to purely geometrical features of space–time, which may also suffice for not taking gravity as something that creates wholes out of parts (if not to the extent that space–time itself also exhibits genuine part–whole relations, which refers us back to our earlier discussion of substantivalism and relationism).

24. One potential difficulty that must be mentioned at this point is that the reality of entanglement has been put into question on the basis of its behaviour under certain conditions. In particular, the phenomenon of entanglement swapping appears problematic in that it consists in entanglement relations being created/destroyed/swapped from one system to another without direct interaction. Indeed, this type of phenomena are among the motivations for non-realist approaches to quantum mechanics as a whole (as in the case of the pragmatist view recently put forward by Healey [2012]). Here, however, in line with Morganti [2009], we will assume that this type of non-realism is not made inevitable (nor particularly compelling) by the relevant evidence about entanglement swapping etc. For discussions, see Cabello [1999], Jordan [1999], Seevinck [2006] and Healey [2012].

25. See also Zimmermann [1996], Schaffer [2003] and Williams [2006].

26. This avoids claiming that only the cosmos truly evolves according to fundamental laws. Schaffer also suggests that the world as a whole is the only fundamental truth-maker [2010b]; and that metaphysical nihilist had better be monists [2007]. In the first case, however, priority monism is assumed rather than argued for, and in the second case the argument is in favour of existence monism, not priority monism. These two arguments, therefore, will not be considered further in what follows.

27. Especially so if one has independent reasons for endorsing a view of quantum mechanics in which it simply makes no sense to speak of the wavefunction of the whole universe.

28. Thinkers such as Leibniz and Whitehead seem to have thought that the universe is (or, at least, might be or might have been) in fact constructed in such a way that everything is a proper part of something.

29. This seems to be in agreement with cosmological theories based on so-called 'chaotic eternal inflation'. See, for instance, Guth and Steinhardt [1984].

30. The idea of an infinite hierarchy of universes, or multiverses, could be regarded as a whimsical hypothesis based on a peculiar reading of the theory

of chaotic eternal inflation. But notice that Schaffer's argument from gunk is primarily based on *a priori* considerations, physics being invoked only to provide secondary (albeit important) external support – the same should then be allowed in the case of arguments from junk.

31. In connection to this, it must be emphasised that the principle of unrestricted composition has been also attacked on further, independent grounds in the past. For instance, on the basis of intuitions about persistence through time, and by claiming that it implies mereological essentialism, leads to paradoxes similar to the ones afflicting naïve set theory, and entails the existence of objects that have properties that are not genuine. Also notice that it is open to discussion whether junk and unrestricted composition (as a necessary truth) are really incompatible. If one thinks they are not, he or she can read the foregoing remarks as directly concerning the mereological postulate of universal sum. For discussion, see Contessa [2012].

32. For discussions of this point, we refer to Cameron [2008], where the idea is put forward that foundationalism should be taken to hold at least contingently, for the actual world in virtue of its explanatory efficacy; and Bohn [2008] and Morganti [2009a], who instead defend the respectability of the denial of foundationalism also for the actual world. Arntzenius and Hawthorne [2005] argued that supporters of the view that the world is gunky need to offer an account of continuous variation. Interestingly in the present context, their proposal as to how to do this is explicitly presented as related to the development of relationism about space and time.

33. For completeness, another issue must be mentioned. Sider [2007] argues that monism (both as priority and as existence monism) is incompatible with our best account of intrinsicality in terms of duplication. Trogdon [2009] suggested an alternative account of intrinsicality, which he then [2010] defended from objections raised by Skiles [2009].

34. In fact, mutual ontic dependence might be regarded as sufficient for elimination: if *x* exists if and only if *y* exists, why not think that, in actual fact, only *x* or only *y* exist, or that *x* and *y* are just aspects of something more fundamental? Indeed, Esfeld and Lam seem to make exactly this step in their latest discussion of structural realism (Esfeld and Lam [2011]), where they suggest that the distinction between objects and structures is merely conceptual.

35. It is ironic that, while they actually use the notion at least implicitly – as it is at the basis of the sort of anti-foundationalism they endorse – Ladyman and Ross make fun of the concept of gunk and of the way it is employed by contemporary analytic metaphysicians [2007; 20]. Strictly speaking, it is true, there is no contradiction between rejecting actual uses of a notion and using it, provided that it is used in a different, novel way; also, Ladyman and Ross explicitly refer to certain uses of the notion of gunk in connection to assumptions about the nature of matter that indeed appear naïve. Nonetheless, the boundary between good and bad uses of metaphysical concepts seems to be far less well-defined than Ladyman and Ross seem to think. At any rate, is its application to contemporary physics and its structural interpretation not already included, at least as a *potentia*, in the concept of gunk as it has been defined by analytic metaphysicians independently of contemporary physics?

36. For a relevant discussion of the metaphysical aspects of ontic structural realism, and a claim to the effect that objects cannot be reduced to structure, and ontological dependence cannot be used to establish strong forms of structural realism, see Wolff [2012].
37. Perhaps, in the sense of something like Paul's [forthcoming[]] mereological bundle theory, which we have mentioned earlier.

Conclusions

Having arrived at the end of the book, it is hoped that readers have the impression that the aims presented at the beginning have at least partly been reached. From the methodological viewpoint, constructive naturalism comes out as a pretty definite (although of course open to criticism) position. The criteria for theory choice being set aside – as they do play an important role, but were not taken to be indicative of anything more than a general view on the interplay between 'sophisticated' and 'common sense' conceptions of the world – it is hoped that the more general approach will be regarded as plausible. As for the case studies, again independently of which positions have been positively argued for in this book, the expectation is that they served to illustrate how constructive naturalism works in practice, and how fruitful the mutual relation between metaphysics and science can be. The specific theses that have been proposed may or may not constitute a coherent whole (for instance, in terms of particulars of some sort, constituting the whole of reality via part–whole relations with various levels of emerging parts and properties, and possibly without a level of entities to be conceived of as fundamental). Be this as it may, further study on both the case studies examined here, as well as others which lend themselves to a treatment in terms of non-reductive metaphysics, cannot but prove philosophically relevant. And even if future results may diverge from those presented here, it is likely that the right approach to these issues goes along the lines identified in this book, with neither elimination nor reduction, but a wise development of metaphysics and science as complementary enterprises – at least as long an ambitious, non-purely-instrumentalist approach to science and our knowledge of reality in general is endorsed.

References

Adams, R.M., (1979): Primitive Thisness and Primitive Identity, *Journal of Philosophy*, 76, 5–25.

Ainsworth, P.M., (2011): Ontic Structural Realism and the Principle of the Identity of Indiscernibles, *Erkenntnis*, 75, 67–84.

Allaire, E.B., (1963): Bare Particulars, *Philosophical Studies*, 14, 1–7.

Allaire, E.B., (1965): Another Look at Bare Particulars, *Philosophical Studies*, 16, 16–20.

Arenhart, J.R.B., (forthcoming): Whither away Individuals, *Synthese*.

Armstrong, D.M., (1997): *A World of States of Affairs*, Cambridge, Cambridge University Press.

Armstrong, D.M., Martin, C.B. and Place, U.T., (1996): *Dispositions: A Debate*, London, Routledge.

Arntzenius, F., (2010): Reichenbach's Common Cause Principle. In E.N. Zalta (ed.), *The Stanford Encyclopedia of Philosophy* (Fall 2010 Edition), URL=<http://plato.stanford.edu/archives/fall2010/entries/physics-Rpcc/>.

Arntzenius, F. and Hawthorne, J., (2005): Gunk and Continuous Variation, *The Monist*, 88, 441–466.

Arntzenius, F. and Maudlin, T., (2010): Time Travel and Modern Physics. In E.N. Zalta (ed.), *The Stanford Encyclopedia of Philosophy* (Spring 2010 Edition), URL=<http://plato.stanford.edu/archives/spr2010/entries/time-travel-phys/>.

Arthur, R., (2006): Minkowski Spacetime and Dimensions of the Present. In Dieks, D. (ed.), *The Ontology of Spacetime*, Amsterdam, Elsevier, 129–155.

Aspect, A., Grangier, P. and Roger, G., (1981): On the Simultaneous Measurement of a Pair of Conjugate Observables, *Bell System Technical Journal*, 44, 725–729.

Aspect, A., Grangier, P. and Roger, G., (1982): Experimental Tests of Realistic Local Theories via Bell's Theorem, *Physical Review Letters*, 47, 460–467.

Audi, P., (2012): A Clarification and Defense of the Notion of Grounding. In Correia, F. and Schnieder, B. (eds), *Metaphysical Grounding*, Cambridge, Cambridge University Press, 101–121.

Auyang, S., (1995): *How is Quantum Field Theory Possible?* Oxford, Oxford University Press.

Ayer, A.J., (1936): Verification and Experience, *Proceedings of the Aristotelian Society*, 37, 137–156.

Bach, A., (1997): *Indistinguishable Classical Particles*, Berlin-Heidelberg, Springer Verlag.

Baia, A., (2012): Presentism and the Grounding of Truth, *Philosophical Studies*, 159, 341–356.

Bain, J., (2000): Against Particle/Field Duality: Asymptotic Particle States and Interpolating Fields in Interacting QFT (Or: Who's Afraid of Haag's Theorem?), *Erkenntnis*, 53, 375–406.

Bain, J., (2011): Quantum Field Theories in Classical Spacetimes and Particles, *Studies in History and Philosophy of Modern Physics*, 42, 98–106.

Baker, D.J., (2009): Against Field Interpretations of Quantum Field Theory, *British Journal for the Philosophy of Science*, 60, 585–609.

Balashov, Y. and Janssen, M., (2003): Presentism and Relativity, *British Journal for the Philosophy of Science*, 54, 327–346.

Ballentine, L.E., (1970): The Statistical Interpretation of Quantum Mechanics, *Reviews of Modern Physics*, 42, 358–381.

Barbour, J., (1999): *The End of Time. The Next Revolution in Our Understanding of the Universe*, London, Weidenfeld and Nicholson.

Barbour, J. and Bertotti, B., (1982): Mach's Principle and the Structure of Dynamical Theories, *Proceedings of the Royal Society*, London, A 382, 295–306.

Barbour, J. and O' Murchada, N., (2010): *Conformal Superspace: The Configuration Space of General Relativity*. Available online at http://arxiv.org/abs/1009.3559.

Barcan Marcus R., (1993): *Modalities: Philosophical Essays*, Oxford, Oxford University Press.

Bardeen, J., Cooper, L.N. and Schrieffer, J.R., (1957): Microscopic Theory of Superconductivity, *Physical Review*, 106, 162–164.

Barnette, R.L., (1978): Does Quantum Mechanics Disprove the Principle of the Identity of the Indiscernibles? *Philosophy of Science*, 45, 466–470.

Baron, S., Evans, P. and Miller, K., (2010): From Timeless Physical Theory to Timelessness, *Humana. Mente*, 13, 35–60.

Bartels, A., (1999): Objects or Events? Towards an Ontology for Quantum Field Theory, *Philosophy of Science*, 66, S170–S184.

Bealer, G., (1999): A Theory of the a Priori, *Philosophical Perspectives*, 13, 29–55.

Bell, J.S., (1964): On the Einstein-Podolsky-Rosen Paradox, *Physics*, 1, 195–200.

Belot, G., (1999): Rehabilitating Relationism, *International Studies in the Philosophy of Science*, 13, 35–52.

Belousek, D.W., (2000): Statistics, Symmetry, and the Conventionality of Indistinguishability in Quantum Mechanics, *Foundations of Physics*, 30, 1–34.

Bigelow, J., (1996): Presentism and Properties, *Philosophical Perspectives*, 10, 35–52.

Bird, A., (2005): The Dispositionalist Conception of Laws, *Foundations of Science*, 10, 353–370.

Bird, A., (2007): *Nature's Metaphysics: Laws and Properties*, Oxford, Oxford University Press.

Black, M., (1952): The Identity of the Indiscernibles, *Mind*, 61, 153–164.

Blackburn, S., (1990): Filling in Space, *Analysis*, 50, 62–63.

Bohn, E.D., (2009): Must There Be a Top Level? *Philosophical Quarterly*, 59, 193–201.

Bourne, C., (2006): *A Future for Presentism*, Oxford, Oxford University Press.

Brighouse, C., (1994): Spacetime and Holes, *Philosophy of Science*, Proceedings 1994, Vol. I, 117–125.

Brighouse, C., (1997): Determinism and Modality, *British Journal for the Philosophy of Science*, 48, 465–481.

Brogaard, B., (2000): Presentist Four-Dimensionalism, *The Monist*, 83, 341–356.

Busch, P., (1999): Unsharp Localization and Causality in Relativistic Quantum Theory, *Journal of Physics A: Mathematics General*, 32, 6535–6546.

Butterfield, J., (1984): Relationism and Possible Worlds, *British Journal for the Philosophy of Science*, 35, 101–113.

Butterfield, J., (1988): Albert Einstein Meets David Lewis, *Philosophy of Science*, Proceedings 1988, Vol. II, 65–81.

Butterfield, J., (1989): The Hole Truth, *British Journal for the Philosophy of Science*, 40, 1–28.

Butterfield, J., (1993): Interpretation and Identity in Quantum Theory, *Studies in history and Philosophy of Science*, 24, 443–476.

Cabello, A., (1999): Quantum Correlations Are Not Local Elements of Reality, *Physical Review A*, 59, 113–115.

Callender, C., (1997): Review of Dorato, M., Time and Reality: Spacetime Physics and the Objectivity of Temporal Becoming, *British Journal for the Philosophy of Science*, 48, 117–120.

Callender, C., (2011): Philosophy of Science and Metaphysics. In French, S. and Saatsi, J. (eds), *The Continuum Companion to the Philosophy of Science*, 33–54.

Calosi, C., Fano, V. and Tarozzi, G., (2011): Quantum Ontology and Extensional Mereology, *Foundations of Physics*, 41, 1740–1755.

Cameron, R.P., (2008): Turtles All the Way Down: Regress, Priority and Fundamentality, *Philosophical Quarterly*, 58, 1–14.

Cameron, R.P., (2011): Truthmaking for Presentists. In Bennett, K. and Zimmerman, D. (eds), *Oxford Studies in Metaphysics*, Vol. 6, Oxford, Oxford University Press, 55–100.

Cameron, R.P., (2012): Composition as Identity Doesn't Settle the Special Composition Question, *Philosophy and Phenomenological Research*, 84, 531–554.

Cao, T.Y., (1997): *Conceptual Developments of Twentieth Century Field Theories*, Cambridge, Cambridge University Press.

Carnap, R., (1950): Empiricism, Semantics, and Ontology, *Revue Internationale De Philosophie*, 4, 20–40.

Casati, R., and Torrengo, G., (2011): The Not so Incredible Shrinking Future, *Analysis*, 71, 240–244.

Caulton, A., (2013): Discerning "Indistinguishable" Quantum Systems, *Philosophy of Science*, 80, 49–72.

Caulton, A. and Butterfield, J., (2012): On Kinds of Indiscernibility in Logic and Metaphysics, *British Journal for the Philosophy of Science*, 63, 27–84.

Chakravartty, A., (2007): *A Metaphysics for Scientific Realism: Knowing the Unobservable*, Cambridge, Cambridge University Press.

Chalmers D.J., (2002): Does Conceivability Entail Possibility? In Gendler, T.S. and Hawthorne, J. (eds), *Conceivability and Possibility*, Oxford, Oxford University Press, 145–200.

Chang, H., and N. Cartwright (1993): Causality and Realism in the EPR Experiment, *Erkenntnis*, 38, 169–190.

Chappell, V.C., (1964): Particulars Re-Clothed, *Philosophical Studies*, 15, 60–64.

Chisholm, R.M., (1990): Events without Times. An Essay on Ontology, *Noûs*, 24, 413–428.

Choi, S. and Fara, M., (2012): Dispositions. In Zalta, E.N. (ed.), *The Stanford Encyclopedia of Philosophy* (Spring 2012 Edition), URL=<http://plato.stanford.edu/archives/spr2012/entries/dispositions/>.

Cleland, C.E., (1984): Space: An Abstract System of Non-Supervenient Relations, *Philosophical Studies*, 46, 19–40.

Clifford, W.K., (1870): *On the Space-Theory of Matter*, read to the Cambridge Philosophical Society on February 21st.

Contessa, G., (2012): The Junk Argument: Safe Disposal Guidelines for Mereological Universalists, *Analysis*, 72, 455–457.

Correia, F., (2008): Ontological Dependence, *Philosophy Compass*, 3, 1013–1032.

Cortes, A., (1976): Leibniz's Principle of the Identity of the Indiscernibles: A False Principle, *Philosophy of Science*, 43, 491–505.

Craig, W.L., (2001): *Time and the Metaphysics of Relativity*, Dordrecht, Kluwer.

Crisp, T., (2004): On Presentism and Triviality. In Zimmerman, D. (ed.), *Oxford Studies in Metaphysics*, Vol. 1, Oxford, Oxford University Press, 15–20.

Crisp, T., (2007): Presentism and the Grounding Objection, *Noûs*, 41, 90–109.

Da Costa, N.C.A. and Krause, D., (1994): Schrödinger Logics, *Studia Logica*, 53, 533–550.

Da Costa, N.C.A. and Krause, D., (1997): An Intensional Schrodinger Logic, *Notre Dame Journal of Formal Logic*, 38, 179–194.

Dalla Chiara, M.L. and Toraldo di Francia, G., (1993): Individuals, Kinds and Names in Physics. In Corsi, G., Dalla Chiara, M.L. and Ghirardi, G. (eds), *Bridging the Gap: Philosophy, Mathematics, Physics*, Dordrecht, Kluwer, 261–283.

Darby, G., (2009): Lewis's Worldmate Relation and the Apparent Failure of Humean Supervenience, *Dialectica*, 63, 195–204.

Darby, G., (2012): Relational Holism and Humean Supervenience, British *Journal for the Philosophy of Science*, 63, 773–788.

Davidson, M., (forthcoming): Presentism and Grounding Past Truths. In Ciuni, R., Torrengo, G. and Miller, K. (eds), *New Papers on the Present: Focus on Presentism*, Munich, Philosophia Verlag.

Dawid, R., (2007): Scientific Realism in the Age of String Theory, *Physics and Philosophy* (online at https://eldorado.tu-dortmund.de/bitstream/2003/24724/1/011.pdf).

De Caro, M. and Macarthur, D., (2012): *Philosophy in an Age of Science. Physics, Mathematics and Skepticism*, Harvard, Harvard University Press.

De Caro, M. and Voltolini, A., (2010): Is Liberal Naturalism Possible? In De Caro, M. and Macarthur, D. (eds), *Normativity and Naturalism*, New York, Columbia University Press, 69–86.

De Clercq, R., (2006): Presentism and the Problem of Cross-Time Relations, *Philosophy and Phenomenological Research*, 72, 386–402.

De Clercq, R., (2012): On Some Putative Graph-Theoretic Counterexamples to the Principle of the Identity of Indiscernibles, *Synthese*, 187, 661–672.

Della Rocca, M., (2005): Two Spheres, Twenty Spheres and the Identity of the Indiscernibles, *Pacific Philosophical Quarterly*, 86, 480–492.

Demirli, S., (2010): Indiscernibility and Bundles in a Structure, *Philosophical Studies*, 7, 1–18.

Diekemper, J., (2009): Thisness and Events, *Journal of Philosophy*, 106, 255–276.

Dieks, D., and Lubberdink, A., (2011): How Classical Particles Emerge from the Quantum World, *Foundations of Physics*, 41, 1051–1064.

Dieks, D. and Versteegh, M., (2008): Identical Quantum Particles and Weak Discernibility, *Foundations of Physics*, 38, 923–934.

Dorato, M., (2006): Absolute Becoming, Relational Becoming and the Arrow of Time: Some Non-Conventional Remarks on the Relationship between Physics and Metaphysics, *Studies in History and Philosophy of Modern Physics*, 37, 559–576.

Dorato, M., (2011): The Alexandroff Present and Minkowski Spacetime: Why It Cannot Do What It Has Been Asked To Do. In Dieks, D., Gonzales, W., Hartmann, S., Uebel, T. and Weber, M. (eds), *Explanation Prediction and Confirmation: New Trends and Old Ones Considered*, Springer, New York, 379–394.

Dorato, M., (2013): *Che Cos'è il Tempo? Einstein, Gödel e l'Esperienza Comune*, Carocci, Rome.

Dorr, C., (2010): Review of Ladyman, J. and Ross, D. (with Spurrett, D. and Collier, J.), Every Thing Must Go: Metaphysics Naturalized, *Notre Dame Philosophical Reviews*, 16 June 2010.

Dyke, H. and Maclaurin, J., (2013): What Shall We Do with Analytic Metaphysics? A Response to McLeod and Parsons, *Australasian Journal of Philosophy*, 91, 179–182.

Earman, J., (1979): Was Leibniz a Relationist? *Midwest Studies in Philosophy*, 4, 263–276.

Earman, J., (2002): Thoroughly Modern McTaggart: Or, What McTaggart Would Have Said If He Had Read the General Theory of Relativity, *Philosophers' Imprint*, 2, 1–28.

Earman, J. and Fraser, D., (2006): Haag's Theorem and its Implications for the Foundations of Quantum Field Theory, *Erkenntnis*, 64, 305–344

Earman, J. and Norton, J., (1987): What Price Spacetime Substantivalism? The Hole Story, *British Journal for the Philosophy of Science*, 38, 515–525.

Einstein, A., Podolski, B. and Rosen, N., (1935): Can Quantum-Mechanical Description of Reality Be Considered Complete? *Physical Review*, 41, 777–780.

Einstein, A., (1948): Quanten-Mechanik und Wirklichkeit, *Dialectica*, 2, 320–324.

Esfeld, M. and Lam, V., (2006): Moderate Structural Realism about Space-Time, *Synthese*, 160, 27–46.

Esfeld, M. and Lam, V., (2011): Ontic Structural Realism as a Metaphysics of Objects. In Bokulich, A., and Bokulich, P. (eds), *Scientific Structuralism*, Dordrecht, Springer, 143–159.

Fine, A., (1984): The Natural Ontological Attitude. In Leplin, J. (ed.), *Scientific Realism*, Berkeley, University of California Press, 83–107.

Fine, A., (1989): Do Correlations Need to Be Explained? In Cushing, J. and McMullin, E. (eds), *Philosophical Consequences of Quantum Theory: Reflections on Bell's Theorem*, Notre Dame, IN, University of Notre Dame Press, 175–194.

Fine, K., (1994): Essence and Modality, *Philosophical Perspectives*, 8, 1–16.

Fine, K., (2002): The Varieties of Necessity. In Szabo Gendler, T. and Hawthorne, J. (eds), *Conceivability and Possibility*, Oxford, Oxford University Press, 253–282.

Fine, K., (2009): The Question of Ontology. In Chalmers, D., Manley, D. and Wasserman, R. (eds), *Metametaphysics: New Essays on the Foundations of Ontology*, Oxford, Oxford University Press, 157–177.

Fine, K., (2010): Some Puzzles of Ground, *Notre Dame Journal of Formal Logic*, 51, 97–118.

Fine, K., (2012): A Guide to Ground. In Correia, F. and Schneider, B. (eds), *Metaphysical Grounding*, Cambridge, Cambridge University Press, 37–80.

Fine, K., (2012a): What is Metaphysics. In Tahko, T.E. (ed.), *Contemporary Aristotelian Metaphysics*, Cambridge, Cambridge University Press, 8–25.

Fiocco, M., (2007): A Defense of Transient Presentism, *American Philosophical Quarterly*, 44, 191–212.

Fleming, G.N., (1989): Lorentz Invariant State Reduction, and Localization, Philosophy of Science, Proceedings 1988, Vol. II, 112–126.

Fleming, G.N. and Butterfield, J., (1999): Strange Positions. In Butterfield, J. and Pagonis, C. (eds), *From Physics to Philosophy*, Cambridge, Cambridge University Press, 108–165.

Fraser, D., (2011): How to Take Particle Physics Seriously: A Further Defence of Axiomatic Quantum Field Theory, *Studies in History and Philosophy of Modern Physics*, 42, 126–135.

French, S., (1989): Individuality, Supervenience and Bell's Theorem, *Philosophical Studies*, 55, 1–22.

French, S., (1995): Hacking Away at the Identity of Indiscernibles: Possible Worlds and Einstein's Principle of Equivalence, *Journal of Philosophy*, 92, 455–466.

French, S., (2003): Scribbling on the Blank Sheet: Eddington's Structuralist Conception of Objects, *Studies in History and Philosophy of Modern Physics*, 34, 227–259.

French, S., (2010): The Interdependence of Structure, Objects and Dependence, *Synthese*, 175, 89–109.

French, S. and Krause, D., (1995): Vague Identity and Quantum Non-Individuality, *Analysis*, 55, 20–26.

French, S. and Krause, D., (2006): *Identity in Physics. A Historical, Philosophical, and Formal Analysis*, Oxford, Oxford University Press.

French, S. and Krause, D., (2010): Remarks on the Theory of Quasi-Sets, *Studia Logica*, 95, 1–2, 101–124.

French, S. and Ladyman, J., (2003): Remodelling Structural Realism: Quantum Physics and the Metaphysics of Structure, *Synthese*, 136, 31–56.

French, S. and Redhead, M., (1988): Quantum Mechanics and the Identity of the Indiscernibles, *British Journal for the Philosophy of Science*, 39, 233–246.

Friedman, M., (1983): *Foundations of Spacetime Theories: Relativistic Physics and Philosophy of Science*, Princeton, Princeton University Press.

Friedman, M., (2001): *Dynamics of Reason. The 1999 Kant Lectures at Stanford University*, Stanford, CSLI Publications.

Gärdenfors, P., (1988): *Knowledge in Flux. Modeling the Dynamics of Epistemic States*, Cambridge, MA, MIT Press.

Gärdenfors, P. and Makinson, D., (1988): *Revisions of Knowledge Systems Using Epistemic Entrenchment*, Proceedings of the Second Conference on Theoretical Aspects of Reasoning about Knowledge, San Francisco, California, 1988. Morgan Kaufmann, 83–95.

Geirsson, H., (2005): Conceivability and Defeasible Modal Justification, *Philosophical Studies*, 122, 279–304.

Gisin, N., (2005): *Can Relativity Be Considered Complete? From Newtonian Nonlocality to Quantum Nonlocality and Beyond*, available online at http://arxiv.org/abs/quant-ph/0512168.

Godfrey-Smith, P., (2006): Theories and Models in Metaphysics, *The Harvard Review of Philosophy*, 14, 4–19.

Gödel, K., (1949): An Example of a New Type of Cosmological Solution of Einstein's Field Equations of Gravitation, *Review of Modern Physics*, 21, 447–450.

Goldman, A.I., (2007): A Program for "Naturalizing" Metaphysics, with Application to the Ontology of Events, *The Monist*, 90, 457–479.

Gottesman, D., (2007): Quantum Statistics with Classical Particles. In Hirota, O., Shapiro, J.H. and Sasaki, M. (eds), *Quantum Communication, Measurement and Computing*, Proceedings of the 8th International Conference on Quantum Communication, Measurement, and Computing, Tokyo, NICT Press, 295–298.

Gracia, J.J.E., (1988): *Individuality: An Essay on the Foundations of Metaphysics*. Albany, New York, State University of New York Press.

Greene, B., (1999): *The Elegant Universe. Superstrings, Hidden Dimensions and the Quest for the Ultimate Theory*, New York, W.W. Norton & Company.

Gryb, S. and Thebault, K., (2012): The Role of Time in Relational Quantum Theories, *Foundations of Physics*, 42, 1210–1238.

Guth, A. and Steinhardt, P. (1984): The Inflationary Universe, *Scientific American*, May, 116–128.

Haag, R., (1955): On Quantum Field Theories, *Matematisk-fysiske Meddelelser*, 29, 1–37.

Hacking, I., (1975): The Identity of Indiscernibles, *Journal of Philosophy*, 72, 249–256.

Hale, B., (1996): Absolute Necessities. In Tomberlin, J.E. (ed.), *Philosophical Perspectives 10: Metaphysics*, Atascadero, CA, Ridgeview, 93–117.

Hales, S.D., (2010): No Time Travel for Presentists, *Logos and Episteme*, 1, 353–360.

Halvorson, H. and Clifton, R., (2002): No Place for Particles in Relativistic Quantum Theories? *Philosophy of Science*, 69, 1–28.

Hansson, S.O., (2011): Logic of Belief Revision. In Zalta, E.N. (ed.), *The Stanford Encyclopedia of Philosophy* (Fall 2011 Edition), URL=http://plato.stanford.edu/archives/fall2011/entries/logic-belief-revision/.

Hawley, K., (2006): Principles of Composition and Criteria of Identity, *Australasian Journal of Philosophy*, 84, 481–493.

Hawley, K., (2009): Identity and Indiscernibility, *Mind*, 118, 101–119.

Healey, R., (1991): Holism and Nonseparability, *Journal of Philosophy*, 88, 393–421.

Healey, R., (2009): Holism and Nonseparability in Physics. In Zalta, E.N. (ed.), *The Stanford Encyclopedia of Philosophy*, Spring 2009 Edition, URL=<http://plato.stanford.edu/archives/spr2009/entries/physics-holism/>.

Healey, R., (2012): Quantum Theory: A Pragmatist Approach, *British Journal for the Philosophy of Science*, 63, 729–771.

Healey, R., (2013): Physical Composition, *Studies in History and Philosophy of Physics*, 44, 48–62.

Hellmann, G., (1982): Stochastic Locality and the Bell Theorems, *Philosophy of Science*, Proceedings 1982, Vol. II, 601–615.

Hilbert, D. and Bernays, P., (1934): *Grundlagen der Mathematik*, Vol. I, Springer, Berlin.

Hill, C., (2006): Modality, Modal Epistemology, and the Metaphysics of Consciousness. In Nichols, S., (ed.), *The Architecture of Imagination*, Oxford, Oxford University Press, 205–236.

Hilborn, R.C. and Yuca, C., (2002): Identical Particles in Quantum Mechanics Revisited, *British Journal for the Philosophy of Science*, 53, 355–389.

Hinchliff, M., (2000): A Defense of Presentism in a Relativistic Setting, *Philosophy of Science*, Proceedings 1998, Vol. II, S575-S586.

Hoefer, C., (1996): The Metaphysics of Space-Time Substantivalism, *Journal of Philosophy*, 93, 5–27.

Hoefer, C., (1998): Absolute versus Relational Spacetime: For Better or Worse, the Debate Goes on', *British Journal for the Philosophy of Science*, 49, 451–467.

Hofweber, T., (2009): Ambitious, Yet Modest, Metaphysics. In Chalmers, D., Manley, D. and Wasserman, R. (eds), *Metametaphysics: New Essays on the Foundations of Ontology*, Oxford, Oxford University Press, 260–289.

Holton, R. (1999): Dispositions All the Way Down, *Analysis*, 59, 9–14.

Horgan, T. and Potrč, M., (2000): Blobjectivism and Indirect Correspondence, *Facta Philosophica*, 2, 249–270.

Horgan, T. and Potrč, M., (2008): Existence Monism Trumps Priority Monism. In Goff, P. (ed.), *Spinoza on Monism*, Houndmills, Palgrave Macmillan.

Howard, D., (1985): Einstein on Locality and Separability, *Studies in History and Philosophy of Science*, 16, 171–201.

Howard, D., (1989): Holism, Separability, and the Metaphysical Implications of the Bell Experiments. In Cushing, J. and E. McMullin, E., (eds), *Philosophical*

Consequences of Quantum Theory: Reflections on Bell's Theorem, Notre Dame, IN, University of Notre Dame Press, 224–253.

Huggett, N., (1995): *What are Quanta, and Why Does it Matter? Philosophy of Science,* Proceedings 1994, Vol. II, 69–76.

Huggett, N., (2003): Quarticles and the Identity of the Indiscernibles. In Brading, K. and Castellani, E. (eds), *Symmetries in Physics: Philosophical Reflections,* Cambridge, Cambridge University Press, 239–249.

Huggett, N. and Norton, J., (forthcoming): Weak Discernibility for Quanta, the Right Way, *British Journal for the Philosophy of Science.*

Huggett, N., Vistarini, T. and Wüthrich, C., (forthcoming): Time in Quantum Gravity. In Bardon, A. and Dyke, H. (eds), *The Blackwell Companion to the Philosophy of Time,* Oxford, Blackwell.

Jackson, F., (1998): *From Metaphysics to Ethics,* Oxford, Oxford University Press.

James, W., (1907[1979]): *Pragmatism, A New Name for Some Old Ways of Thinking, Popular Lectures on Philosophy,* New York Longmans, Green and Company.

Jarrett, J.P., (1984): On the Physical Significance of the Locality Conditions in the Bell Arguments, *Noûs,* 18, 569–589.

Jarrett, J.P., (1989): Bell's Theorem: A Guide to the Implications. In Cushing, J. and McMullin, E. (eds), *Philosophical Consequences of Quantum Theory: Reflections on Bell's Theorem,* University of Notre Dame Press, Notre Dame, IN, 60–79.

Jeshion, R., (2006): The Identity of the Indiscernibles and the Co-Location Problem, *Pacific Philosophical Quarterly,* 87, 163–176.

Jones, M. and Clifton, R., (1993): Against Experimental Metaphysics. In French, P., Euling, J.T.E. and Wettstein, H. (eds), *Midwest Studies in Philosophy,* Vol. XVIII, University of Notre Dame Press, Notre Dame, 295–316.

Jordan, T.F., (1999): Quantum Correlations Violate Einstein-Podolsky-Rosen Assumptions, *Physical Review A,* 60, 2726–2728.

Kantorovich, A., (2003): The Priority of Internal Symmetries in Particle Physics, *Studies in History and Philosophy of Modern Physics,* 34, 651–675.

Keller, S., and Nelson, M., (2011): Presentists Should Believe in Time-Travel, *Australasian Journal of Philosophy,* 79, 333–345.

Krause, D., (1992): On a Quasi-set Theory, *Notre Dame Journal of Formal Logic,* 33, 402–411.

Kriegel, U., (2008) Composition as a Secondary Quality, *Pacific Philosophical Quarterly,* 89, 359–383.

Kriegel, U., (2012): Kantian Monism, *Philosophical Papers,* 41, 23–56.

Kuhlmann, M., (2010): *The Ultimate Constituents of the Material World – In Search of an Ontology for Fundamental Physics,* Frankfurt, Ontos Verlag.

Ladyman, J., (1998): What is Structural Realism? *Studies in History and Philosophy of Science,* 29, 409–424.

Ladyman, J., (2007): *On the Identity and Diversity of Individuals,* Proceedings of the Aristotelian Society, Supplementary Volume, 81, 23–43.

Ladyman, J., (2009): Structural Realism. In Zalta, E.N. (ed.), *Stanford Encyclopedia of Philosophy,* Summer 2009 Edition, URL=<http://plato.stanford.edu/archives/sum2009/entries/structural-realism/>.

Ladyman, J., (2012): Science, Metaphysics and Method, *Philosophical Studies,* 160, 31–51.

Ladyman, J. and Bigaj, T., (2010): The Principle of the Identity of Indiscernibles and Quantum Mechanics, *Philosophy of Science,* 77, 117–136.

Ladyman, J., and Leitgeb, H., (2008): Criteria of Identity and Structuralist Ontology, *Philosophia Mathematica,* 16, 388–396.

Ladyman, J. and Ross, D. (with Spurrett, D. and Collier, J.) (2007): *Every Thing Must Go. Metaphysics Naturalised*, Oxford, Oxford University Press.

Ladyman, J., Linnebo, O. and Pettigrew, R., (2012): Identity and Discernibility in Philosophy and Logic, *Review of Symbolic Logic*, 5, 162–186.

Lange, M., (2002): *An Introduction to the Philosophy of Physics: Locality, Fields, Energy, and Mass*, London, Wiley-Blackwell.

Laudan, L., (1981): A Confutation of Convergent Realism, *Philosophy of Science*, 48, 19–49.

Leeds, S., (2007): Physical and Metaphysical Necessity, *Pacific Philosophical Quarterly*, 88, 458–485.

Lehmkuhl, D., (2011): Mass-Energy-Momentum: Only There Because of Spacetime? *British Journal for the Philosophy of Science*, 62, 453–488.

Leibniz, G.W., (1704[1956]): *The Leibniz-Clarke Correspondence*, edited by H.G. Alexander, Manchester University Press, Manchester.

Leibniz, G.W., (1704[1981]): *Nouveau Essais sur L'Entendement Humain*, originally published in 1765. Translated as *New Essays on Human Understanding* by Remnant, P. and Bennett, J., Cambridge, Cambridge University Press.

LePoidevin, R., (1991): *Change, Cause and Contradiction*, London, MacMillan.

Lewis, D., (1986): *On the Plurality of Worlds*, London, Wiley-Blackwell.

Lewis, D., (1991): *Parts of Classes*, London, Wiley-Blackwell.

Lewis, D., (1994): Humean Supervenience Debugged, *Mind*, 103, 473–490.

Loewer, B., (2004): Humean Supervenience. In Carroll, J. (ed.), *Readings on Laws of Nature*, Pittsburgh, University of Pittsburgh Press, 176–206.

Lowe, E.J., (2001): *The Possibility of Metaphysics: Substance, Identity, and Time*, Oxford, Oxford University Press.

Lowe, E.J., (2003): Individuation. In Loux, M.J. and Zimmerman, D.W. (eds), *The Oxford Handbook of Metaphysics*, Oxford, Oxford University Press, 75–95.

Lowe, E.J., (2010): Ontological Dependence. In Zalta, E.N. (ed.), *The Stanford Encyclopedia of Philosophy*, Spring 2010 Edition, URL=<http://plato.stanford.edu/archives/spr2010/entries/dependence-ontological/>.

Lowe, E.J., (2011): The Rationality of Metaphysics, *Synthese*, 178, 99–109.

Lowe, E.J., (forthcoming): Grasp of Essences versus Intuitions: An Unequal Contest. In Booth, T. and Rowbottom, D. (eds), *Intuitions*, Oxford, Oxford University Press.

Lupher, T., (2010): Not Particles, Not Quite Fields: An Ontology for Quantum Field Theory, *Humana. Mente*, 13, 155–174.

Lyre, H., (2004): Holism and Structuralism in U(1) Gauge Theory, *Studies in History and Philosophy of Modern Physics*, 35, 643–670.

MacDonald, A., (2001): Einstein's Hole Argument, *American Journal of Physics*, 69, 223–225.

Mackie, J.L., (1973): *Truth, Probability and Paradox*, Oxford, Oxford University Press.

Mackie, J.L., (1977): Dispositions, Grounds and Causes, *Synthese*, 34, 361–370.

Mackie, P., (2006): *How Things Might Have Been: Individuals, Kinds, and Essential Properties*, Oxford, Oxford University Press.

Maclaurin, J. and Dyke, H., (2012): What is Analytic Metaphysics for?, *Australasian Journal of Philosophy*, 90, 291–306.

Maddy, P., (2007): *Second Philosophy: A Naturalistic Method*, Oxford, Oxford University Press.

Malament, D., (1996): In Defence of Dogma – Why There Cannot Be a Relativistic Quantum Mechanical Theory of (Localizable) Particles. In Clifton, R. (ed.), *Perspectives on Quantum Reality: Non-Relativistic, Relativistic, and Field-Theoretic*, Dordrecht and Boston, Kluwer, 1–10.

Margenau, H., (1944): The Exclusion Principle and its Philosophical Importance, *Philosophy of Science*, 11, 187–208.

Markosian, N., (1998): Brutal Composition, *Philosophical Studies*, 92, 211–249.

Markosian, N., (1998a): Simples, *Australasian Journal of Philosophy*, 76, 213–228.

Massimi, M., (2001): Exclusion Principle and the Identity of the Indiscernibles: A Response to Margenau's Argument, *British Journal for the Philosophy of Science*, 52, 303–330.

Maudlin, T., (1994): *Quantum Nonlocality and Relativity*, Oxford, Blackwell.

Maudlin, T., (2002): Thoroughly Muddled McTaggart: Or, How to Abuse Gauge Freedom to Create Metaphysical Monstrosities, *Philosophers' Imprint*, 2, 1–23.

Maudlin, T., (2007): *The Metaphysics within Physics*, Oxford, Oxford University Press.

Maxwell, N., (1988): Quantum Propensiton Theory: A Testable Resolution of the Wave/Particle Dilemma, *British Journal for the Philosophy of Science*, 39, 1–50.

McCall, S., (1994): *A Model of the Universe*, Oxford, Oxford University Press.

McDaniel, K., (2003): Against Maxcon Simples, *Australasian Journal of Philosophy*, 81, 265 – 275.

McDaniel, K., (2008): Against Composition as Identity, *Analysis*, 68, 128–133.

McLeod, M. and Parsons, J., (2012): Maclaurin and Dyke on Analytic Metaphysics, *Australasian Journal of Philosophy*, 91, 173–178.

McTaggart, J.M.E., (1908): The Unreality of Time, *Mind*, 17, 457–474.

McTaggart, J.M.E., (1927): *The Nature of Existence*, Vol. II, Cambridge, Cambridge University Press.

Meiland, J.W., (1966): Do Relations Individuate? *Philosophical Studies*, 17, 65–69.

Meiland, J.W., (1974): A Two-Dimensional Passage Model of Time for Time Travel, *Philosophical Studies*, 26, 153–173.

Meyer, U., (2005): The Presentist's Dilemma, *Philosophical Studies*, 122, 213–225.

Miller, K., (2005): What is Metaphysical Equivalence? *Philosophical Papers*, 34, 45–74.

Minkowski, H., (1907): Das Relativitätsprinzip, *Annalen der Physik*, 352, 927–938.

Misner, C.W. and Wheeler, J.A., (1957): Classical Physics as Geometry: Gravitation, Electromagnetism, Unquantized Charge, and Mass as Properties of Curved Empty Space, *Annals of Physics*, 2, 525–603.

Monton, B., (2006): Presentism and Quantum Gravity. In Dieks, D. (ed.), *The Ontology of Spacetime*, Dordrecht, Kluwer, 263–280.

Monton, B., (2011): Prolegomena to Any Future Physics-Based Metaphysics, *Oxford Studies in Philosophy of Religion*, Vol. III, 142–165.

Morganti, M., (2004): On the Preferability of Epistemic Structural Realism, *Synthese*, 142, 81–107.

Morganti, M., (2009): *A New Look at Relational Holism in Quantum Mechanics*, *Philosophy of Science*, Proceedings 2008, 1027–1038.

Morganti, M., (2009a): Ontological Priority, Fundamentality and Monism, *Dialectica*, 63, 271–288.

Morganti, M., (2011): The Partial Identity Account of Partial Similarity Revisited, *Philosophia*, 39, 527–546.

Muller, F.A., (2011): How to Defeat Wüthrich's Abysmal Embarrassment Argument against Space-Time Structuralism, *Philosophy of Science*, Proceedings 2010, 1046–1057.

Muller, F.A., (2011a): Whithering Away, Weakly, *Synthese*, 180, 223–233.

Muller, F.A. and Linnebo, O., (forthcoming): *On Witness Discernibility*, Erkenntnis.

Muller, F.A. and Saunders, S., (2008): Discerning Fermions, *British Journal for the Philosophy of Science*, 59, 499–548.

Muller, F.A. and Seevinck, M.P., (2009): Discerning Elementary Particles, *Philosophy of Science*, 76, 179–200.

Mumford, S., (1998): *Dispositions*; Oxford, Oxford University Press.

Mumford, S., (2004): *Laws in Nature*, London, Routledge.

Ney, A., (2012): Neo-Positivist Metaphysics, *Philosophical Studies*, 160, 53–78.

Neta, R., (2007): Review of De Caro, M. and Macarthur, D. (eds), Naturalism in Question, *The Philosophical Review*, 116, 657–663.

Newton, I., (1687[1999]): *Philosophiae Naturalis Principia Mathematica*, translated as *Mathematical Principles of Natural Philosophy* by Cohen, I.B., Whitman, A. and Budenz, J., Berkeley, University of California Press.

Newton-Smith, W.H., (1980): *The Structure of Time*, London, Routledge and Kegan Paul.

Noonan, H., (2013): Presentism and Eternalism, *Erkenntnis*, 78, 219–227.

Oderberg, D., (2011): Essence and Properties, *Erkenntnis*, 75, 85–111.

Oppenheim, P. and Putnam, H., (1958): The Unity of Science as a Working Hypothesis. In Feigl H., Scriven, M. and Maxwell, G. (eds), *Concepts, Theories and the Mind-Body Problem, Minnesota Studies in Philosophy of Science*, Minneapolis, University of Minnesota Press, Vol. II, 3–36

Paul, L.A., (2012): Metaphysics as Modeling: The Handmaiden's Tale, *Philosophical Studies*, 160, 1–29.

Paul, L.A., (2012a): Building the World from Its Fundamental Constituents, *Philosophical Studies*, 158, 221–256.

Paul, L.A., (forthcoming): Mereological Bundle Theory. In Burkhardt, H., Seibt, J. and Imaguire, G. (eds), *Handbook of Mereology*, Munich, Philosophia Verlag.

Peacocke, C., (1998): *Being Known*, Oxford, Oxford University Press.

Peirce, C.S., (1905): Issues of Pragmaticism, *The Monist*, 15, 481–499.

Peterson, D. and Silberstein, M., (2010): Relativity of Simultaneity and Eternalism: In Defense of Blockworld. In Petkov, V. (ed.), *Space, Time, and Spacetime: Physical and Philosophical Implications of Minkowski's Unification of Space and Time*, Berlin, Springer.

Pianesi, F. and Varzi, A.C., (1996): Events, Topology and Temporal Relations, *The Monist*, 79, 89–116.

Pooley, O., (2001): *Relationalism Rehabilitated? II: Relativity*. Available online at http://philsci-archive.pitt.edu/221/1/rehab2ps.pdf.

Pooley, O., (forthcoming): Substantivalist and Relationalist Approaches to Spacetime. In Batterman, R. (ed.), *The Oxford Handbook of Philosophy of Physics*, Oxford, Oxford University Press.

Pooley, O. and Brown, H.R., (2002): Relationalism Rehabilitated? I: Classical Mechanics, *British Journal for the Philosophy of Science*, 53, 183–204.

Popper, K.R., (1957): The Propensity Interpretation of the Calculus of Probability and of the Quantum Theory. In Körner, S. (ed.), *Observation and Interpretation. A Symposium of Philosophers and Physicists*, New York, Dover, Buttersworth Scientific Publications, 65–70.

Prior, E.W., Pargetter R. and Jackson, F., (1982): Functionalism and Type-Type Identity Theories, *Philosophical Studies*, 42, 209–225.

Putnam, H., (1967): Time and Physical Geometry, *Journal of Philosophy*, 64, 240–247.

Putnam, H., (1975): *Mathematics, Matter and Method*, Cambridge, Cambridge University Press.

Putnam, H., (2004): *Ethics without Ontology*, Cambridge, Harvard University Press.

Quine, W.v.O., (1951): On Carnap's Views on Ontology, *Philosophical Studies*, 2, 65–72.

Quine, W.v.O., (1960): *Word and Object*, Cambridge, MA, MIT Press.

Quine, W.v.O., (1976): Grades of Discriminability, *Journal of Philosophy*, 73, 113–116.

Quine, W.v.O. and Ullian, J.S., (1978): *The Web of Belief*, New York, McGraw-Hill.

Ramsey, F.P., (1925): Universals, *Mind*, 34, 401–417.

Rasmussen, J., (2012): Presentists May Say Goodbye to A-properties, *Analysis*, 72, 270–276.

Raven, M.J., (2012): In Defence of Ground, *Australasian Journal of Philosophy*, 90, 687–701.

Redhead, M., (1982): Quantum Field Theory for Philosophers, *Philosophy of Science*, Proceedings 1982, Vol. II, 57–99.

Redhead, M., (1995): More Ado About Nothing, *Foundations of Physics*, 25, 123–137.

Redhead, M., (1996): *From Physics to Metaphysics*, Cambridge, Cambridge University Press.

Redhead, M. and Teller, P., (1992): Particle Labels and the Theory of Indistinguishable Particles in Quantum Mechanics, *British Journal for the Philosophy of Science*, 43, 201–218.

Reeh, H. and Schlieder, S., (1961): Bermerkungen zur Unitäräquivalenz von Lorentz-invarianten Feldern, *Nuovo Cimento*, 22, 1051–1068.

Reichenbach, H., (1956): *The Direction of Time*, Berkeley, University of Los Angeles Press.

Reid, T., (1785[2002]): *Essays on the Intellectual Powers of Man*, edited by Brookes, D., University Park, Pennsylvania State University Press.

Rietdijk, C.W., (1966): A Rigorous Proof of Determinism Derived from the Special Theory of Relativity, *Philosophy of Science*, 33, 341–344.

Ritchie, J., (2008): *Understanding Naturalism*, Durham, Acumen.

Rodriguez-Pereyra, G., (2004): The Bundle Theory is Compatible with Distinct but Indiscernible Particulars, *Analysis*, 64, 72–81.

Rosen, G., (forthcoming): Metaphysical Dependence: Grounding and Reduction'. In Hale, B. and Hoffman, A. (eds), *Modality: Metaphysics, Logic, and Epistemology*, Oxford, Oxford University Press, 109–136.

Rosen, G. and Dorr, C., (2002): Composition as a Fiction. In Gale, R. (ed.), *The Blackwell Companion to Metaphysics*, Oxford, Blackwell.

Rott, H., (2001): *Change, Choice and Inference. A Study of Belief Revision and Nonmonotonic Reasoning*, Oxford, Clarendon Press.

Rovelli, C., 2009): Forget Time, written for the FQXi 'The Nature of Time' essay contest. Http://fr.arxiv.org/pdf/0903.3832v3.

Russell, B., (1911): On the Relations of Universals and Particulars. In Marsh, R.C. (ed.), *Logic and Knowledge: Essays 1901 to 1950*, London, Allen and Unwin, 105–124.

Russell, B., (1914): *Our Knowledge of the External World as a Field for Scientific Method in Philosophy*, London, George Allen and Unwin.

Rynasiewicz, R., (1994): The Lessons of the Hole Argument, *British Journal for the Philosophy of Science*, 45, 407–436.

Rynasiewicz, R., (1996): Absolute Versus Relational Space-Time: An Outmoded Debate? *Journal of Philosophy*, 93, 279–306.

Rynasiewicz, R., (2011): Newton's Views on Space, Time, and Motion. In Zalta, E.N. (ed.), *The Stanford Encyclopedia of Philosophy*, Fall 2011 Edition, URL=<http://plato.stanford.edu/archives/fall2011/entries/newton-stm/>.

San Pedro, I., (2011): Causation, Measurement Relevance and No-conspiracy in EPR, *European Journal for the Philosophy of Science*, 2, 137–156.

Saunders, S., (1994): A Dissolution of the Problem of Locality, *Philosophy of Science*, Proceedings 1994, Vol. II, 88–98.

Saunders, S., (2003): Physics and Leibniz's Principles. In Brading, K. and Castellani, E. (eds), *Symmetries in Physics: Philosophical Reflections*, Cambridge, Cambridge University Press, 289–308.

Saunders, S., (2006): Are Quantum Particles Objects? *Analysis*, 66, 52–63.

Saunders, S., (2006a): On the Explanation for Quantum Statistics, *Studies in History and Philosophy of Modern Physics*, 37, 192–211.

Savitt, S., (2000): There's No Time Like the Present (in Minkowski Spacetime), *Philosophy of Science*, 67, Proceedings 1998, vol. II, S563–S574.

Savitt, S., (2009): The Transient Nows. In Myrvold, W.C. and Christian, J. (eds), *Quantum Reality, Relativistic Causality, and Closing the Epistemic Circle*, The Western Ontario Series in Philosophy of Science 74, Amsterdam, Springer, 339–352.

Schaffer, J., (2003): Is There a Fundamental Level? *Noûs*, 37, 498–517.

Schaffer, J., (2007): From Nihilism to Monism, *Australasian Journal of Philosophy*, 85, 175–191.

Schaffer, J., (2009): On What Grounds What. In Manley, D., Chalmers, D. and Wasserman, R. (eds), *Metametaphysics: New Essays on the Foundations of Ontology*, Oxford, Oxford University Press, 347–383.

Schaffer, J., (2009a): Spacetime the One Substance, *Philosophical Studies*, 145, 131–148.

Schaffer, J., (2010): Monism: The Priority of the Whole, *Philosophical Review*, 119, 31–76.

Schaffer, J., (2010a): The Internal Relatedness of All Things, *Mind*, 119, 341–376.

Schaffer, J., (2010b): The Least Discerning and Most Promiscuous Truthmaker, *Philosophical Quarterly*, 60, 307–324.

Schlick, M., (1978): Philosophical Papers I, 1909–1922, Vienna Circle Collection 11, Heath, P. (trans.), Mulder, H.L. and Van de Velde-Schlick, B.F.B. (eds), Dordrecht, Reidel.

Seevinck, M.P., (2006): The Quantum World Is Not Built Up from Correlations, *Foundations of Physics*, 36, 1573–1586.

Seibt, J., (2002): Quanta, Tropes or Processes: Ontologies for QFT Beyond the Myth of Substance. In Kuhlmann, M., Lyre, F. and Wayne, A. (eds), *Ontological Aspects of Quantum Field Theory*, Singapore, World Scientific, Singapore.

Sellars, W., (1962): Philosophy and the Scientific Image of Man. In Colodny, R. (ed.), *Frontiers of Science and Philosophy*, Pittsburgh, University of Pittsburgh Press, 35–78.

Shalkowski, S., (1997): Essentialism and Absolute Necessity, *Acta Analytica*, 12, 41–56.

Shimony, A., (1980): Critique of the Papers of Fine and Suppes, *Philosophy of Science*, Proceedings 1980, Vol. II, 572–580.

Shoemaker, S., (1969): Time without Change, *Journal of Philosophy*, 66, 363–381.

Shoemaker, S., (1980): Causality and Properties. In Van Inwagen, P. (ed.), *Time and Cause: Essays Presented to Richard Taylor*, Dordrecht, Reidel, 109–135.

Shoemaker, S., (1998): Causal and Metaphysical Necessity, *Pacific Philosophical Quarterly*, 79, 59–77.

Sider, T., (2001): *Four Dimensionalism: An Ontology of Persistence and Time*, Oxford, Oxford University Press.

Sider, T., (2006): Bare Particulars, *Philosophical Perspectives*, 20, 387–397.

Sider, T., (2007): Against Monism, *Analysis*, 67, 1–7.

Sider, T., (forthcoming): Consequences of Collapse. In Baxter, D. and Cotnoir, A. (eds), *Composition as Identity*, Oxford, Oxford University Press.

Simons, P., (2002): Candidate General Ontologies for Situating Quantum Field Theory. In Kuhlmann, M., Lyre, F. and Wayne, A. (eds), *Ontological Aspects of Quantum Field Theory*, Singapore, World Scientific, 33–52.

Skiles, A., (2009): Trogdon on Monism and Intrinsicality, *Australasian Journal of Philosophy*, 87, 149–154.

Skow, B., (2005): *Once Upon a Spacetime*, PhD Dissertation, New York University.

Skow, B., (2009): Relativity and the Moving Spotlight, *Journal of Philosophy*, 106, 666–678.

Smeenk, C. and Wüthrich, C., (2011): Time Travel and Time Machines. In Callender, C. (ed.), *The Oxford Handbook of Philosophy of Time*, Oxford, Oxford University Press.

Smith, Q., (1996): The Metaphysical Necessity of Natural Laws, *Proceedings of the Heraclitean Society*, 18, 104–123.

Stachel, J., (2002): 'The Relations between Things' versus 'the Things between Relations': The Deeper meaning of the Hole Argument. In Malament, D. (ed.), *Reading Natural Philosophy: Essays in the History and Philosophy of Science and Mathematics*, Chicago and LaSalle, Open Court, 231–266.

Stapp, H., (1979): Whiteheadian Approach to Quantum Theory and the Generalized Bell's Theorem, *Foundations of Physics*, 9, 1–25.

Stein, H., (1970): On the Notion of Field in Newton, Maxwell, and Beyond. In Stuewer, R.B. (ed.), *Historical and Philosophical Perspectives of Science*, Minnesota Studies in the Philosophy of Science, Vol. V, Minneapolis, University of Minnesota Press, 264–287.

Stoljar, D., (2010): *Physicalism*, Milton Park and New York, Routledge.

Stoneham, T., (2009): Time and Truth: The Presentism-Eternalism Debate, *Philosophy*, 84, 201–218.

Strawson, G., (2008): The Identity of the Categorical and the Dispositional, *Analysis*, 68, 271–282.

Strawson, P.F., (1959): *Individuals. An Essay in Descriptive Metaphysics*, Routledge, London.

Suarez, M., (2007): Quantum Propensities, *Studies in History and Philosophy of Modern Physics*, 38, 418–438.

Swoyer, C., (1982): The Nature of Natural Laws, *Australasian Journal of Philosophy*, 60, 203–223.

Tahko, T.E., (2011): A Priori and A Posteriori: A Bootstrapping Relationship, *Metaphysica*, 12, 151–164.

Tahko, T.E., (2012): *Counterfactuals and Modal Epistemology*, Grazer Philosophische Studien, 86, 93–115.

Tallant, J., (forthcoming): Intuitions in Physics, *Synthese*.

Tallant, J., (forthcoming a): Defining Existence Presentism, *Erkenntnis*.

Tallant, J., (forthcoming b): Quantitative Parsimony and the Metaphysics of Time: Motivating Presentism, *Philosophy and Phenomenological Research*.

Tallant, J., (2012): (Existence) Presentism and the A-theory, *Analysis*, 72, 673–681.

Tallant, J., (2009): Presentism and Truth-Making, *Erkenntnis*, 71, 407–416.

Tallant, J., and Ingram, D., (2012): Time for Distribution? *Analysis*, 72, 264–270.

Tegmark, M., (2007): The Mathematical Universe, *Foundations of Physics*, 38, 101–150.

Teller, P., (1986): Relational Holism and Quantum Mechanics, *British Journal for the Philosophy of Science*, 37, 71–81.

Teller, P., (1990): Prolegomenon to a Proper Interpretation of Quantum Field Theory, *Philosophy of Science*, 57, 594–618.

Teller, P., (1995): *An Interpretive Introduction to Quantum Field Theory*, Princeton, Princeton University Press.

Teller, P. and Redhead, M., (2000): Is Indistinguishability in Quantum Mechanics Conventional? *Foundations of Physics*, 30, 951–957.

Tersoff, J. and Bayer, D., (1983): Quantum Statistics for Distinguishable Particles, *Physical Review Letters*, 50, 553–554.

Trogdon, K., (forthcoming): An Introduction to Grounding. In Hoeltje, M., Schnieder, B. and Steinberg, A. (eds), *Dependence*, Munich, Philosophia Verlag.

Trogdon, K., (2010): Intrinsicality for Monists (and Pluralists), *Australasian Journal of Philosophy*, 88, 555–558.

Trogdon, K., (2009): Monism and Intrinsicality, *Australasian Journal of Philosophy*, 87, 127–148.

Van Fraassen, B., (1991): *Quantum Mechanics: An Empiricist View*, Oxford, Clarendon.

Van Fraassen, B., (2002): *The Empirical Stance*, New Haven, Yale University Press.

Van Inwagen, P., (1990): *Material Beings*, Ithaca, NY, Cornell University Press.

Van Inwagen, P., (2012): Metaphysics. In Zalta, E.N. (ed.), *The Stanford Encyclopedia of Philosophy*, Winter 2012 Edition, URL=http://plato.stanford.edu/archives/win2012/entries/metaphysics/.

Van Kampen, N., (1984): The Gibbs Paradox. In Parry, W.E. (ed.), *Essays in Theoretical Physics in Honour of Dirk ter Haar*, Oxford, Pergamon Press, 303–312.

Wallace, D., (2011): Taking Particle Physics Seriously: A Critique of the Algebraic Approach to Quantum Field Theory, *Studies in History and Philosophy of Modern Physics*, 42, 116–125.

Wallace, D. and Timpson, C.G., (2010): Quantum Mechanics on Spacetime I: Spacetime State Realism, *British Journal for the Philosophy of Science*, 61, 697–727.

Warmbröd, K., (2004): Temporal Vacua, *Philosophical Quarterly*, 54, 266–286.

Weingard, R., (1972): Relativity and the Reality of Past and Future Events, *British Journal for the Philosophy of Science*, 23, 119–121.

Weyl, H., (1927[1949]). *Philosophie der Matemathik und Naturwissenschaft*, Munich, Oldenbourg, English translation *Philosophy of Mathematics and Natural Science*, Princeton, Princeton University Press.

Whitehead, A.N., (1929[1979]): *Process and Reality: An Essay in Cosmology*, corrected edition by Griffin, D.R. and Sherburne, D.W., New York, Free Press.

Williams, J.R.G., (2006): Illusions of Gunk, *Philosophical Perspectives*, 20, 493–513.

Williamson, T., (2007): Philosophical Knowledge and Knowledge of Counterfactuals, *Grazer Philosophiche Studien*, 74, 89–123.

Winsberg, E., and Fine, A., (2003): Quantum Life: Interaction, Entanglement and Separation, *Journal of Philosophy*, 100, 80–97.

Wolff, J., (2012): Do Objects Depend on Structures? *British Journal for the Philosophy of Science*, 63, 607–625.

Worrall, J., (1989): Structural Realism: the Best of Both Worlds? *Dialectica*, 43, 99–124.

Wüthrich, C., (2010): Demarcating Presentism. In De Regt, H., Okasha, S. and Hartmann, S. (eds), *EPSA Philosophy of Science Amsterdam 2009*, Dordrecht, Springer, 439–448.

Wüthrich, C., (2010a): No Presentism in Quantum Gravity. In Petkov, V. (ed.), *Space, Time, and Spacetime: Physical and Philosophical Implications of Minkowski's Unification of Space and Time*, Berlin, Springer, 257–278.

Yablo, S., (1993): Is Conceivability a Guide to Possibility? *Philosophy and Phenomenological Research*, 53, 1–42.

Zeh, H., (2004): The Wave-function: It or Bit? In Barrow, J., Davies, P. and Harper, C. (eds), *Science and Ultimate Reality: Quantum Theory, Cosmology and Complexity*, Cambridge, Cambridge University Press, 103–120.

Index

Printed in the United States
by Baker & Taylor Publisher Services